EVALUATION RESEARCH ON THE BENEFITS OF GRAZING SYSTEM IN THE MONGOLIAN GRASSLAND

蒙古草原放牧制度效益评价研究

那音太　秦福莹　田志馥 ◎ 著

本书获得国家自然科学基金项目"基于高分光学与高分极化SAR的蒙古高原湿地退化指示种识别研究"（61661045）的资助

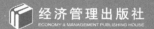

图书在版编目（CIP）数据

蒙古草原放牧制度效益评价研究/那音太，秦福莹，田志馥著. —北京：经济管理出版社，2021.6

ISBN 978-7-5096-8044-5

Ⅰ.①蒙⋯　Ⅱ.①那⋯　②秦⋯　③田⋯　Ⅲ.①放牧管理—评价—内蒙古　Ⅳ.①S815.2

中国版本图书馆 CIP 数据核字（2021）第 106602 号

组稿编辑：王光艳
责任编辑：高　娅
责任印制：黄章平
责任校对：王淑卿

出版发行：经济管理出版社
　　　　　（北京市海淀区北蜂窝 8 号中雅大厦 A 座 11 层　100038）
网　　址：www.E-mp.com.cn
电　　话：（010）51915602
印　　刷：北京晨旭印刷厂
经　　销：新华书店
开　　本：720mm×1000mm/16
印　　张：11.5
字　　数：177 千字
版　　次：2021 年 7 月第 1 版　2021 年 7 月第 1 次印刷
书　　号：ISBN 978-7-5096-8044-5
定　　价：68.00 元

·版权所有　翻印必究·
凡购本社图书，如有印装错误，由本社读者服务部负责调换。
联系地址：北京阜外月坛北小街 2 号
电话：（010）68022974　　邮编：100836

Preface

Grasslands are not only the main source of feed for livestock but also have the ability to conserve water, regulate atmosphere and water cycle, improve ecological environment functions and values, as well as provide habitat for some wildlife. For these reasons, grasslands are a key component of the Earth's terrestrial ecosystem. In recent years, there is an obvious trend of grassland degradation worldwide due to the combined effects of climate change and human activities. Widespread land desertification, water resources depletion, species extinction, food shortage and other severe ecological and socio-economic issues have deteriorated the entire ecosystem. Hence, it has caught the attention of countries around the world, and the Mongolian Plateau is a typical example.

The Mongolian Plateau consists of vast areas of grasslands, dominated by typical grasslands, in a transition zone between arid and semi-arid climates. With an extremely fragile ecosystem where its main economic activity is grassland husbandry, it is extremely sensitive to climate change and external disturbances. The Mongolian Plateau is an important ecological barrier in northern China, and its ecological destruction and degradation will have an important impact on China's ecological security. Meanwhile, the Mongolian Plateau is also one of the most important regions for China's national development

蒙古草原放牧制度效益评价研究
Evaluation Research on the Benefits of Grazing System in the Mongolian Grassland

strategy of One Belt One Road (OBOR), and the construction of the China – Mongolia – Russia economic corridor. Nevertheless, in the last half – century, the rate, scale and extent of grassland degradation have been greater than at any time in history. Desertification has a tendency to spread to large areas, and even to degrade. For instance, in the past 50 years, meadow grasslands, typical grasslands and desert grasslands have declined in vegetation productivity by 54% ~ 70%, 30% ~ 40% and 50% respectively. The balance and stability of the grassland ecosystem has been severely disturbed and disrupted. Hence significantly reduced its functionality, increasing severity of various natural disasters, and causing great damage to the local economy and social development. For these reasons, an in – depth study of the causes of ecological changes in the region is of strategic and practical importance for ecological security, social stability and economic development in China, Mongolia, Russia, Asia and even the world.

Currently, there are many academic and social perspectives on the causes of the rapid degradation of the Mongolian steppes, particularly the Inner Mongolian steppes. In addition to global climate change, the dominant view is "overgrazing". Therefore, the implementation of "grass and livestock balance" and "emigration" are proposed to achieve a harmonious relationship between people and pastoral land. It is noteworthy that the Mongolian plateau, an extremely fragile ecosystem with a history of thousands of years of harmonious human – nature development, has been severely destructed in the modern era, and that the Mongolia's steppes are much better than those of the Inner Mongolia Autonomous Region. Is this related to differences in management practices? Therefore, theoretical breakthroughs may be obtained from a comparative study of the grassland environment under different grazing regimes in Mongolia and Inner Mongolia Autonomous Region.

This book selected two adjacent soums (one on each side of the international border) adjacent to each other in the China – Mongolia typical steppes transboundary area with basically the same natural conditions as the study area. Combining the quadrat sampling method and remote sensing, we set up three perpendicular transects that dissected the two countries' international border and seven transects parallel to the international border to the response of plant communities to the grazing system in the natural state. Then, in combination with remote sensing data and questionnaires, we discussed the causes of Mongolian steppes degradation from the perspective of grazing systems.

The book is divided into six chapters: The first chapter is the introduction of the book, which mainly provides an overview of the grassland, a review of grazing ecology researches, and the background and significance of this research; the second chapter is about the study location, which focuses on the evolution of grazing systems in the Mongolian plateau; the third chapter is about the study design and method; the fourth chapter is the analysis of result; the fifth chapter is the discussion; and the sixth chapter is the conclusion.

This book was funded by the National Natural Science Foundation of China (NSFC) for the study, "Identification Study of the Mongolian Plateau Wetlands degradation indicator species based on the high resolution Optical and polarization SAR images" (61661045).

Dr. NA Yintai

March, 2021

Contents

Chapter 1 Introduction 1

  1.1 An Overview of Grasslands 1

  1.2 Study Overview 11

    1.2.1 The Study of Grassland Grazing Ecosystem 11

    1.2.2 The Study of Grazing Systems 17

    1.2.3 The Study on the Effect of Grazing on Grassland Communities 24

    1.2.4 The Study on the Effect of Grazing Systems on Grassland Communities in Mongolia 27

  1.3 Research Background and Significance 32

    1.3.1 Research Background 32

    1.3.2 Research Significance 33

  1.4 Research Content and Purpose 34

    1.4.1 Research Content 34

    1.4.2 Research Purpose 35

Chapter 2   Study Area ·· 36

2.1   An Overview of Mongolian Plateau ·· 36
    2.1.1   Natural Condition ·· 36
    2.1.2   An Overview of Socio – Economic ·· 42

2.2   The Evolution of Grazing System ·· 53
    2.2.1   The Evolution of Grazing System in Mongolia ·· 53
    2.2.2   The Evolution of Grazing System in Inner Mongolia Autonomous Region ·· 56
    2.2.3   Comparison of Grazing Systems in China and Mongolia ············ 63

2.3   The Overview of the Study Site ·· 64
    2.3.1   Sukhbaatar Province ·· 66
    2.3.2   The Abag Banner of Xilingol League ·· 67

Chapter 3   Research Design and Methodology ·· 70

3.1   Research Methodology ·· 70
    3.1.1   Field Survey Method ·· 70
    3.1.2   Remote Sensing Monitoring Method ·· 71
    3.1.3   Questionnaire Survey Method ·· 72
    3.1.4   Historical Documentation Method ·· 72
    3.1.5   Statistical Analysis Method ·· 72

3.2   Experimental Setup ·· 73

3.3   Quadrat Survey ·· 74
    3.3.1   General Characteristics of Plants ·· 74
    3.3.2   Calculation of Species Dominance ·· 75

3.3.3 Calculation of Species Diversity ········ 75

3.3.4 Classification of Plant Groups ········ 77

3.3.5 Normalized Difference Vegetation Index (NDVI) ········ 77

3.3.6 Calculation of Plant Community Stability ········ 79

Chapter 4 Analysis of Results ········ 83

4.1 General Characteristics of Plant Community ········ 83

4.2 Analysis of Species Dominance ········ 85

4.3 Analysis of Species Diversity ········ 89

4.4 Characterization of Plant Functional Groups ········ 92

4.5 Normalized Difference Vegetation Index (NDVI) ········ 95

4.6 Plant Community Stability Index ········ 98

Chapter 5 Discussion ········ 101

5.1 Effects of Different Grazing Systems on Plant Community Characteristics ········ 101

5.2 Effects of Different Grazing Systems on the Dominance of Major Species ········ 103

5.3 Effects of Different Grazing Systems on Species Diversity ········ 109

5.4 Effects of Different Grazing Systems on Plant Functional Groups ········ 116

5.5 Effects of Different Grazing Systems on NDVI ········ 122

5.6 Effects of Different Grazing Systems on Plant Community Stability ········ 125

Chapter 6　Conclusion ………………………………………………… 130

References ……………………………………………………………… 135

Appendix ………………………………………………………………… 171

Chapter 1 Introduction

1.1 An Overview of Grasslands

　　Broadly speaking, grasslands are the largest ecosystem in the world, covering an area of approximately 52.5 million km^2, or 40.5% of the land area outside of Greenland and Antarctica. In a narrower sense, the grasslands are a type of land cover with herbaceous plants as the dominant species, and with either sparse trees or no trees at all. UNESCO defines land with tree and shrub coverage of less than 10% as grassland, and 10% ~40% as wooded grassland (Suttie et al., 2011). The term "grassland", is known by many names around the world. For example, it is called "prairie" in North America, "Pampas" in South America, "Velde" and "Savanna" in Africa, and "Steppe" in Asia and Europe. Although there are many definitions for grasslands, most refer to the UNESCO and Oxford Dictionary: The grasslands consist of herbaceous vegetation formed in more arid environments (Berretta et al., 2005).

蒙古草原放牧制度效益评价研究
Evaluation Research on the Benefits of Grazing System in the Mongolian Grassland

With the continuous development of human society, humans have gradually evolved from the primitive natural resources collectors to agricultural producers (the primary industry) and then to industry developers (secondary industry). In recent times, with the increasing industrialization and Informationization of human society, the proportion of traditional agriculture and livestock production in the society's gross output value has gradually declined. Meanwhile, the proportion of the gross output value of the tertiary industry such as services and information industry shows an increasing trend. Nevertheless, in the progress of human society development, agriculture and livestock production are still essential because food and nutrition are the basic needs of human survival. Therefore, the primary industry has always been the most critical factor in determining the development of human society in all times and contexts. Grasslands are an important natural resource on which livestock production is based, and their role in supporting human social and economic development has never been weakened.

Grasslands are not only the main source of feed for livestock but also have the ability to conserve water, regulate atmosphere and water cycle, improve ecological environment functions and values, as well as provide habitat for some wildlife. For these reasons, grasslands are a key component of the Earth's terrestrial ecosystem. The grassland ecosystem not only has economic value which can support the development of livestock production, but also is essential in protecting terrestrial ecosystems and environment. Desertification of grasslands and the generation of dust storms are concrete manifestations of the degradation of the grassland's ecological barrier, which is related to the scientific aspect of the utilization and management of grasslands (Li, 1997). The natural grassland ecosystem of Mongolian Plateau constitutes a natural barrier for ecological security in Northeast Asia. If this grassland ecological system is destroyed, it will severely threaten the ecological security of Northeast Asia and even the entire Asia (Qin, 2019). In

Chapter 1　Introduction

China, the steppe of Inner Mongolia has important economic value and ecological value. The importance of grassland ecosystems in Inner Mongolia as an important barrier to protect the ecological security of northern China is gradually being appreciated. For the last 30 years, the global trend of grassland degradation is significant, and the overall level of ecological environment tends to deteriorate, hence attracting the attention of all countries in the world (Liu et al., 2005). The main reason for the serious damage to the grassland ecosystem in Inner Mongolia is the unscientific use of the steppe, of which the root cause is the conflict of interests, which must be given sufficient attention (Wang et al., 1996; Na et al., 2010).

The health of the natural grassland ecosystem is affected by various causes. For example, the fire, whether wild or man-caused fire, are constantly and extensively affecting the grassland (Humphrey, 1974). Another main cause is the foraging of livestock and wildlife (Han & Rithchie, 1998; Li & Wang, 1999; Li & Li, 2002). Other causes include logging to clear an area for cultivation, fencing and water point construction, and a range of "improvement measures" such as harrowing and replanting (Wu, 2002; Jia, 2011). Semple (1971) used to summarize practical technologies in grassland utilization, most of which are relevant to current technologies and issues, with some technologies only progressing in detail. In general, if the grassland is not cultivated or seeded, the grassland ecosystem is natural.

Of course, many of the long cultivated grasslands are no longer relevant to the seeding at the time of establishment. A large area of grasslands with better water conditions in the world has been reclaimed as farmland, especially the North America prairies, South American Pampas, and Eurasian steppes. Grazing is carried out on marginal lands unsuitable for cultivation, and in these areas people's basic means of subsistence are mainly provided by domestic animals (Burke et al., 1989; Li et al., 2012; Li et al.,

2007). In many places, only a small part of the grassland has not been reclaimed because their precipitation does not meet the needs of crop growth (Zhao et al., 2002). Clearing grasslands for crops has a significant impact on grasslands, such as reducing large areas of grazing space, obstructing traditional migration routes for livestock movement, and preventing livestock from reaching watering points (Ren et al., 2000; Cao et al., 2006).

Although large areas of grassland are being utilized globally for cultivation or other uses, there are still vast areas of grazing land, including the steppes of Mongolia and the northern part of China, which extending into Europe (the area of grasslands in the north of China has declined considerably compared to the last 100 years though), the Tibetan Plateau and its adjacent Hindu Kush Himalayan Region's mountain pasture, the North American prairies, South American Pampas, the tropical Campos grasslands, South American grasslands and savannas, Patagonian grasslands and plateaus, Australian grasslands, Mediterranean and West Asia's large areas of semi – arid grazing land, the vast Sahel and Sudanese Gizahel regions in the southern Sahara desert, and much of eastern Africa from the Cape of Good Hope to the Mozambique Channel (Suttie et al., 2011).

The grazing system can be roughly divided into two types, the commercial and traditional. The traditional type mainly aims to sustain life. The use of extensive grazing is a major form of natural grassland ecosystem management. This non – crop land use is in competition with other land uses such as crops, wildlife, forests and recreation. The use of grassland is not fixed but depends on economic, soil, climate and other influencing factors. The grassland, which is restricted by topography, poor soil, and growing season, is generally not suitable for intensive farming but for grazing production. (Bao & Hai, 1995). In many countries, grazing grasslands can also include arid, gravelly, floodplain, mountainous or remote regions. Therefore, all discussion about grasslands

Chapter 1 Introduction

must take place in the context of animal production, and human livelihoods (Baker, 1993).

Commercial grazing operations on natural pastures are usually larger in scale and are generally dominated by a single species of livestock, such as beef cattle or sheep for wool. Since the 19th century, large-scale grazing systems have been gradually developed on pastures with low grazing livestock density or ungrazed pastures, mainly due to migration from the Americas and Oceania, while the level of development in Africa and Asia is lower (Wario et al., 2016; Mccall & Clark, 1999; Dyer, 2002).

Traditional livestock production systems vary according to climate and type of regional agricultural system. Such a system also raises a wide variety of domestic livestock, including horses, goats, yaks, camels, etc., in addition to cattle and sheep. Domestic livestock are often used for a variety of purposes, such as meat, milk, fibers, service force, and their dung for fuels. In many national and regional cultures, the number of domestic livestock is related to the status of their owner.

Traditional livestock production systems are mostly distributed in the agricultural-pastoral areas, and usually use a crop-livestock farming system because livestock use straw, crop by-products, and also grassland that is not suitable for crops-growing. However, the vast grasslands are still used as nomadic or seasonal grazing grounds, with herds moving seasonally among grazing sites according to season, and in some cases according to temperature, or according to the availability of forage for emergency use. Other factors can also affect livestock migration. For example in the Great Lakes Basin region of Mongolia, pastoralists have to leave the lower grazing areas near Uvs Lake in June due to insect bites that spread disease and enter the mountains, returning only in the fall; but then they have to migrate to the mountains again in the winter to avoid the threat to the survival of grazing livestock posed by the extreme cold temperatures in the basin (Li, 1994; Erdenebaatar, 2003).

The use of the terms "nomadic grazing" and "seasonal grazing" are sometimes confused in the context of migratory livestock production systems. In reality, however, seasonal grazing is a grazing system in which people and livestock move between two distinctly different seasonal pastures that are far apart or have large differences in elevation. Nomadic grazing refers to herds that graze without fixed location and move with precipitation (Koch et al., 2017). The herds of nomadic grazing often include a wide variety of livestock, which helps reduce risk while taking advantage of the vegetation – different species of animals can feed on different plants (Zhang et al., 2007). In the past 150 years, political and economic changes in the world have had a significant impact on the distribution, production status and use of grasslands, with an unprecedented increase in settlement and the reclamation of grasslands for crop cultivation. With the independence of colonial countries, the democratization of the authoritarian state usually led to the loss of traditional authority and pastoral land rights but the problem of private livestock grazing on public grasslands remains. The management system of communal grazing grasslands in Central Asian countries, China and Russia was transformed from feudal systems to collective management, but the development of collective economies gradually waned in the following two decades. The way in which grasslands are managed and the patterns of their use vary from country to countries. For example, after the collapse of the Soviet Union, Russia placed state – owned land under the management of collective farms and state farms, which in turn distributed the land to farm members in the form of nominal land rights (Shvidenko & Nilsson, 2010).

In the 1950s and 1960s, the Soviet Union carried out a massive reclamation campaign in arable areas at the expense of the grasslands. The first crop after reclamation was so productive that most of the remaining grassland was quickly reclaimed for cultiva-

Chapter 1 Introduction

tion and almost never sown pasture. Perhaps they did not realize that fertility accumulated gradually through grasslands and did not consider that this practice caused soil degradation (Shan et al., 1996).

In the 1970s and 1980s, direct grazing of livestock using grasslands became rare in the Soviet Union, replaced by the use of large – scale farms and zero grazing, which was based on forage crops such as corn and oats. The implementation of the measures was mostly concentrated in the extreme periphery of mixed farms, ignoring the idea of a crop – livestock farming system. Compared to past decades, village herds are increasing and are beginning to be raised freely around the countryside, and communal or public grazing resources are increasingly being threatened by the privatization of livestock ownership. A viable solution is to give assistance to weaker pastures, livestock, crops and soils, especially for small mixed family farms. While the earlier large Soviet collective farm – style non – land units are retained as the central and collective core, livestock production will still shift to a family – based approach. Therefore, family farms have to rely on their farm products and land they operate to feed their livestock. This provides a solid basis for farm – grassland crop rotation (Lerman, 2013).

In recent years, due to the multiple impacts of global climate change and human activities, a series of serious ecological, environmental and socio – economic problems such as land desertification, water depletion, species extinction and food shortage have emerged in different degrees around the world. Among them, desertification is one of the typical manifestations of global ecological degradation. By 2010, more than 1 billion people in more than 100 countries and regions were harmed by desertification and desertified land accounted for about 1/3 of the total global land area, causing the world to suffer the most direct economic loss of $ 423 million per year (Li, 2011). The deteriorating ecological environment has severely limited the sustainable development of regional economies, such as the Sahara Desert in Africa, the Mongolian Plateau, and terrestrial ecosystems in West Asia, which are facing enormous challenges (Fang, 2000; Zhuo, 2007).

The Mongolian steppe is one of the most important parts of the temperate Eurasian steppe (Bao et al., 2010), and most of its areas belong to arid and semi – arid zones, with little rainfall and a very fragile ecological environment, which is very sensitive to environmental changes. Mongolia and Inner Mongolia of China are sharing the main part of the Mongolian steppes, which have been under pressure from the most intense human economic activities for the last half century. The impact of human interference and global climate change on the degradation of Mongolian grasslands has become increasingly serious, thus drawing the attention of scholars working on climate, hydrology, and ecology (Angerer et al., 2008; Yun et al., 2011; Endicott, 2003). Mongolia is one of the few purely pastoralist countries, and its population and livestock numbers have grown rapidly over the past half century. Although overgrazing exists, the grassland ecosystem remains in good condition to a considerable extent, compared to the more prominent degradation of grasslands in China (Conte & Tilt, 2014).

The famous grassland of Horqin at the beginning of the 20th century changed over a century to the present Horqin sandy land, which is a typical case of grassland desertification caused by human factors (Na, 2010; Zhao & Zhou, 1994; Wu, 2002).

Dr. Na and his colleagues addressed the lack of quantitative information on the relationship between Mongolian steppe degradation and population for medium and long term (50 ~ 300 years). Using Horqin grassland in Inner Mongolia as the study area, they used linear regression analysis and literature review methods to explore the characteristics and causes of population change in Horqin. A correlation analysis between the proportion of degraded grassland area and population density was carried out using 5 pe-

Chapter 1 Introduction

riods of TM remote sensing data from 1977 ~ 2014 in a typical geographical area to quantify the effect of population change on grassland degradation. The results show that immigration and immigrant fertility are the largest factors contributing to population growth in the region, which shows a high rate of population growth under the influence of immigration and fertility encouragement policies. There is a positive correlation between the proportion of degraded grassland area and its population density in the two typical regions, i. e. , grassland degradation tends to be more severe as the population density increases (Na et al. , 2010) . With the growing population, the huge demand for food, clothing and other basic elements of survival and livelihood has put heavy pressure on land and water resources, placing higher demands on the carrying capacity of these resources. In order to meet the large demand for household and industrial products, people have overexploited natural resources such as mineral deposits, forests and grasslands, thus further accelerating the process of grassland degradation (Hu, 2000; Jin et al. , 2007; Xie & Peng, 2017) .

A number of professionals and scholars have suggested that excessive ecological pressure on the population and its continued growth are the root causes of the deterioration of the grassland ecosystem in Inner Mongolia. The population density in pastoral areas of northern and northwestern China (including Inner Mongolia) is much lower than in the eastern and southern region. However, basic natural features such as drought (precipitation of only 200 ~ 300 mm) and the rapid population growth after 1949 have put increasing pressure on the grasslands, which are the main natural resource available to pastoral areas, thus ultimately putting increasing pressure on the potential for sustainable development of pastoral areas in this fragile ecological environment (John et al. , 1994) . In 2016, Bilige and Du also affirmed the theory that demographic factors led to the degradation of grasslands in Inner Mongolia. They pointed out that the common points of several organized immigrations after 1949 were the migration of people from the interior to the Inner Mongolia Autonomous Region and their engagement in agricultural production activities. Besides, the reclamation of grasslands and the shrinking of their areas have resulted in limited space for livestock development, desertification, and serious ecological damage. Aorenqi (2005) suggested that the main causes of grassland degradation in Inner Mongolia are firstly, the change of pastoralists' lifestyle from nomadism to settlement, secondly, the unreasonable fencing, and finally, the reclamation of grasslands or the introduction of agricultural production model, by which he meant the change of nomadism to settlement, the reclamation of grasslands and the introduction of agricultural production model are all subsidiary phenomena of population growth. In the early 1990s, the carrying capacity was overshot by 84% in China. Calculating based on this number while taking population growth into account, the national population was expected to peak in the 20th century, even if the population growth rate in pastoral land was 30% of the national average for the same period. Based on the average annual increase in living standards of farmers and pastoralists of only 2% (assuming that this improvement is obtained entirely by local farming and animal husbandry), the grassland carrying capacity may be very high, up to more than 300% (Hou, 2001). It can be seen that solving the problem of overgrazing and pasture degradation in the grasslands has put forward very high requirements on the speed and magnitude of the transfer of the grassland farming and herding population and farming labor (emigrant or industrial restructuring) in the coming decades.

Chapter 1　Introduction

1.2　Study Overview

1.2.1　The Study of Grassland Grazing Ecosystem

The concept of economic management of grassland ecosystems is derived from the early establishment and development of Clements – Duksterhuis succession theory (Li, 1993; Yang, 2008). The theory of equilibrium was developed from the early scientific study of grasslands in North America and is represented by Clements's Plant Succession, published in 1916, in which he proposed a concept of community succession that was of great importance in guiding a series of related studies conducted later in the field of ecology (Glenn – Lewin et al., 1993).

In 1919, Sampson was the first to apply equilibrium theory to the practice of range management. Although the theory is based on a specific, small – scale, wet climate case study of the Nebraska Great Plains, the theory's creators believe it still has some general applicability (Sampson, 1919). In 1949, Dyksterhuis conducted a systematic analysis and summary of the division of natural grazing land conditions and trends in balanced ecosystems, which laid a solid foundation for the large – scale application of balanced ecosystem theory in the management of natural grazing land. The theory had been applied satisfactorily in practice of natural grasslands management in the United States where rainfall is abundant and relatively stable. By the 1950s, the theory of plant community succession proposed by Clements gradually became one of the important theories guiding the study of plant ecology. The practical activities of pasture management using equilibrium theory cover the following aspects: Changing from nomadic to settlement; setting fences and defining the boundaries of pastoral land rights; enforcing pasture stocking rate; replacing subsistence livestock production with market – oriented livestock production; improving livestock breeds, etc. The above model is gradually adopted and applied to the management practice of grazing pastures all over the world, and its importance and authority even exceeds that of the ecological theory itself. In the ensuing period, the theory was introduced into social, economic, cultural, and political fields and had a profound impact on the development of countries around the world. In Clements – Duksterhuis succession theory, it is assumed that the grazing system, which has deviated from equilibrium because of disturbance, will automatically return to its original state or enter a new equilibrium after the disturbance is removed (Wu, 2017). It has been argued that the principle of equilibrium ecosystem is applicable to typical grassland grazing systems where the degree of degradation is relatively less severe. However, a significant number of studies on grassland grazing ecosystems have questioned the generality of their use, especially the results of researches conducted in more fragile ecosystems could not provide valid support for the theory. In order to further revealing the mechanisms that produce fluctuating trends in grassland ecosystems and explain the stability and complexity of grazing systems, the constraints of equilibrium ecosystem theory were abandoned and some scholars introduced the non – equilibrium theory. In fact, for grazing systems with a very high degree of degradation, it follows the principle of non – equilibrium theory.

Since the late 1970s, the field of ecological research has generally recognized that the theory of dynamic equilibrium is difficult to achieve in many ecosystems and that it has relatively obvious limitations: ①This theory suggests that vegetation response to grazing pressure in a linear or reversible manner, and therefore vegetation condition can

Chapter 1 Introduction

be regulated by controlling stocking rate; ② Adjustment of the number of grazing animals using a density-driven model; ③ Abiotic factors do not affect vegetation change, etc.

The results of two studies conducted in the African region showed that: Density-driven model is not applicable to arid and semi-arid regions with erratic and scarce rainfall; many specific grassland ecosystems are also characterized by neither absolute equilibrium nor absolute non-equilibrium, but rather exhibit a combination of both (Wyatt-Smith, 1987; Coppock et al., 1988). Traditional equilibrium theory assumes the existence of a steady state, with the central idea that every system is at some steady state, whereby the system can deviate from its equilibrium point and fluctuate up and down within a certain range. Pickett and his colleagues (1994) pointed out that a balanced system should have the following four characteristics: The system is enclosed; possesses some self-regulation ability; has a stable equilibrium state and a defined direction of change, and is not subjected to external disturbances. In his study of the Yellowstone National Park landscape ecosystem in the United States, Romme found that the park's plant community composition and diversity were in a process of continuous, significant fluctuation, and therefore concluded that the landscape ecosystem was a non-equilibrium system. The theory of non-equilibrium systems believes that the ecosystem is open, and the regulation of internal factors tends to shift to a stable state, but when the system is disturbed by external factors, there will be certain fluctuations. The uncertainty in external disturbances is an important reason for the unstable performance of the system, even though the system has an internally stable equilibrium point, the direction of its change and succession presents uncertainty. When external disturbances are considered as part of the system, the system exhibits complexity and its uncertainty, nonlinearity and integration are further enhanced. Assuming that there is an equilibrium state in an ecosystem, that equilibrium state is only a state at a particular spatial and temporal scale. Therefore, in terms of the fundamental characteristics of ecosystems, they remain open, non-equilibrium, uncertain and complex. Besides, when conducting research on ecosystem theory, simple linear relationships to describe complex ecosystems should be avoided. Moreover, the non-equilibrium theory argues that a series of preconditions are required for the classical ecological equilibrium theory to be established, and for a period of time, the equilibrium theory has been widely and wrongly adopted worldwide as a commonly used theory. The use of flawed theories to guide ecosystem research and practical activities has also created many problems. Currently, the application of non-equilibrium theory-based perspectives in grassland ecological research and pasture management is attracting widespread attention (Romme, 1982).

For typical grasslands formed under temperate inland semi-arid climatic conditions, dry and widespread perennial grasses dominate their vegetation communities, hence provide a comprehensive characteristics and profile of temperate grasslands (Li, 1988). When competition occurs between plant populations, stress from environmental factors, disturbance from grazing activities and the synergistic, complex effect of both are the main disturbance factors driving changes in vegetation. In the study of dynamic characteristics of grassland ecosystems, drought is a key factor affecting the stability of grassland ecosystems. Semi-arid grazing system is located near the center of the competition-stress-disturbance (CSP) triangle (Grime, 1979). Although the three factors are important to the system, none can have an overwhelming influence on the system (Milton, 1994). Although the relative equilibrium relationships among the above three factors vary somewhat among plant communities, the process of elucidating the complexity and stability of semi-arid grazing systems has led to the derivation of equilibrium and non-equilibrium ecological principles of grazing systems, which are opposing views.

Chapter 1　Introduction

Xiong and his colleagues summarized in quite details, the researches carried out by predecessors on the dynamic characteristics of typical grassland ecosystems in the Xilin River basin. They pointed out that typical grassland ecosystems exhibit equilibrium and non - equilibrium characteristics, and depending on grazing intensity, degradation and other conditions, grassland ecosystems show fluctuating succession trends. First, the theory of equilibrium ecosystems is applicable for typical grassland grazing ecosystems with less degradation (Xiong et al., 2004). The Clements - Duksterhuis succession view based on the principle of equilibrium ecosystem is an important theoretical basis for conducting studies on the dynamics of typical grassland grazing systems in the Xilin River basin, but the previously conducted studies on the grazing systems in the Xilin River basin mainly focused on less degraded grasslands, which was consistent with the reality of grassland grazing utilization in the region at that time. Second, the theory of equilibrium ecosystem is not applicable to the more degraded ecosystems (Westoby et al., 1989). In the past 10 years or so, due to the dramatic increase of grassland stocking rate, the degradation of grassland has been increasingly aggravated and the area of severely degraded grassland has been expanding under the dual effect of two factors: Continuous overgrazing and arid climatic conditions. One of the direct results of this is the gradual dominance of a pioneer leguminous shrub species, *Caragana microphylla*, turning a typical grassland into scrubland, which has become a widespread phenomenon throughout the Xilin River basin at present. Given the current situation of severe degradation and transitioning of grasslands in the basin into scrubland, it is clear that the classical succession view of Clements - Duksterhuis based on the equilibrium principle can no longer provide an adequate and reasonable explanation, so it is urgent to introduce a non - equilibrium ecosystem perspective. This view is of great practical significance for promoting the in - depth study of grassland grazing systems in the Xilin River basin. Moreover, under the guidance of the new theory, it facilitates the scientific implementation of practical activities to restore degraded grassland ecosystems. Lastly, the principle of non – equilibrium ecosystem applies to more severely degraded ecosystems. The principle of non – equilibrium ecosystem states that in grassland grazing systems, stochastic events determine the structure of the vegetation and the composition of the community. Abiotic variables such as rainfall (rainfall intensity and its seasonal distribution) can have a decisive influence on vegetation dynamics and thus on changes in herbivore communities. In this non – equilibrium system, the effective water supply becomes the main factor driving the changes in the system. "Event – driven" system fluctuations, i. e. the randomness and probability of abiotic (e. g. precipitation) and biotic (e. g. grazing) factors determine vegetation dynamics, plant and animal populations fluctuate over a wide range, and the system exhibits obvious non – equilibrium characteristics. When the complexity of the system increases, the ability of humans to control the system reduces. The state and excess model based on the principle of non – equilibrium ecosystem, systematically characterizes various alternate states of vegetation by virtue of "event – driven", where the main variables that cause changes in vegetation state include: climate change, grazing or scrub removal, etc. When the disturbance is removed, the system may not be restored to its original state. In a balanced grazing system, the community succession, i. e., shifts between vegetation states, are not always reversible. Under the influence of disturbance factors, vegetation community succession may occur with the disappearance of certain species, which in turn may cause irreversible changes in interspecific community relationships or significant environmental changes, and therefore, irreversible community succession may occur. Some of the transitions that occur in the process result in changes in the relative strength of facilitation and competition due to the loss of species and their functions in the system. The shift be-

Chapter 1 Introduction

tween reversible and irreversible changes is closely related to limit value – the threshold of the intensity of the impact of each ecological factor. When the value is within threshold value, the changes are reversible and the system returns to its original state; when the threshold value is exceeded, the changes are irreversible and the system cannot return to its original state.

From its own point of view, the grassland ecosystem is subject to many uncertainties. The grassland ecosystem is an important foundation for the survival of human society, and human social activities and cultural concepts are decisive factors for the health and stability of the grassland ecosystem. In the past, among the studies on the dynamic change pattern of grassland ecosystem and grassland management and utilization, ecologists have mostly focused on ecological balance and conservation, while economists have mainly considered economic benefits, and studies taking both into account have not been common. However, an increasing number of studies are starting to replace single – idea – guided studies with a sustainable development concept that focuses on both ecology and economy. Meanwhile, the construction of ecological civilization in China has been brought to a strategic level, and the healthy environment has been paid more and more attention, which provides a strong guarantee for the in – depth study of the evolution of grassland ecosystems and scientific utilization methods.

1.2.2 The Study of Grazing Systems

The economic activities involved in grassland ecosystems are dominated by livestock, for which the grazing system plays a decisive role (Wang & Chen, 1999). The grazing system is the system of organization and utilization in grazing management, which systematically arranges the time and space utilization of livestock grazing land. The grazing system provides a scientific combination of grass utilization and grooming in

蒙古草原放牧制度效益评价研究
Evaluation Research on the Benefits of Grazing System in the Mongolian Grassland

the temporal and spatial dimensions, combining grazing intensity and grazing system adjustments to achieve a quantitative balance between pasture growth and livestock nutritional requirements (Vera, 1991).

Grazing systems and methods are differentiated depending on the natural conditions of the pasture, the type of grass, the season, the type of livestock, and the grazing habits. Each grazing method has its own specific historical background and conditions of formation, and each has its own characteristics to meet the different needs of livestock production. The grazing system is actually a matter of rational and effective use of grassland resources in terms of spatial and temporal allocation. Grazing systems vary according to the focus of the classification criteria, and their terminology also differs. According to the different ways of organization and management, the grazing system can generally be divided into two categories: Continuous grazing system and zoned rotational grazing system (Wu, 2017).

Continuous grazing system includes single – species grazing system and multi – species grazing system. Continuous grazing system refers to the utilization of a pasture throughout the year. The system can save cost, and increase livestock exercise volume and enhance their immune system compared to raising in shelter. Moreover, light grazing can increase the productivity of pasture and plant community diversity, and help improve soil structure and composition. In contrast, when the degree of grazing is higher than the carrying capacity of the pasture, pasture degradation will occur. Zoned rotational grazing is to divide a seasonal grazing land or year – round grazing land into several divisions according to the productivity of grass and the number of grazing livestock, and grazing activities occur within each division for a number of days before shifting to another. Several to dozens of rotational grazing divisions are calculated as a unit, and each unit is used by a group of livestock, where they graze area by area in a rotation. Zoned

Chapter 1   Introduction

rotational grazing includes seasonal camp grazing, fenced grazing and confinement feeding and grazing. Seasonal camp grazing is the division of pasture into seasonal camps, each of which is utilized for a certain grazing season. Confinement feeding and grazing refers to the production method mainly focusing on confinement feeding, and supplemented by grazing. Zoned rotational grazing's advantages mainly include the following: First, it can reduce the waste of pasture resources and save the area of grass being utilized. Second, it can optimize the vegetation composition structure and improve the yield and quality of forage. Rotational grazing can evenly utilize grassland, prevent weed growth, and promote the growth of good forage grasses, while free grazing is mostly detrimental to the growth of good forage grasses and leads to the deterioration of vegetation structure. Third, it can increase production of livestock. Rotational grazing can prevent livestock from consuming too much thermal energy due to activities, which is conducive to healthy growth and development. Fourth, by strengthening the management of grazing land, rotational grazing facilitates the planned implementation of appropriate agro – technical measures. Fifth, it is to prevent the transmission of parasitic helminths among livestock. Generally, rotational grazing methods are not grazed in the same pasture for more than six consecutive days, which can reduce the chance of livestock infected with larvae.

Since the 1950s, some countries and regions, including the United States, have developed a variety of special grazing systems based on zoned rotational grazing, including deferred rotation grazing, rest rotation grazing, seasonal grazing, high – intensity low – frequency grazing (HILF), short duration grazing (SDG), optimal grazing land grazing and Merrill grazing system (Three Herd, Four Pasture), etc. The HILF grazing system, also known as high – intensity utilization grazing or non – selective grazing, was widely used in the 1960s. Based on the HILF grazing system, Booysen proposed a short duration grazing system in the late 1960s (Chang & Xia, 1994). Under this system, with high stock density and shortened grazing periods, livestock are often able to feed on fresh pasture, resulting in a significant increase in daily diet quality. Under better management conditions, the stocking rate is substantially higher than that of free grazing and other grazing systems, hence this grazing system is promoted in all grassland types around the world (Orr et al., 2003). Kothmann divided the special grazing systems into four mutually distinct groups based on the summary of the effectiveness of various grazing systems applied in grassland production. They are deferred rotation grazing, rest rotation grazing, HILF and SDG (China Agricultural University as the chief editor, 1982). Reasonable grazing system can help restore the vitality of grassland, improve the productivity of grassland, maintain the ecological balance of grassland, and enable the sustainable use of grassland (Romme, 1982; Li, 1988; Pickett, 1994).

Research on grazing systems outside China has a history of more than 100 years, and the system of zoned rotational grazing in grassland first originated in Western Europe. The term rotational grazing was first described in the "Agronomist – Farmer's Dictionary" published by French scientists in 1760, followed by studies in England, the Netherlands and other countries. In 1887, South Africa began to advocate a zoned rotational grazing system (Armour et al., 1994). In 1895, the American scholar Jared Smith suggested that the condition of the grasslands in the southern Great Plains of the United States could be improved by zoned rotational grazing, and his idea was supported by Smapson, Jardine and Anderson in their respective experiments conducted in different areas (Cheng et al., 2014). In recent years, foreign scholars have started to conduct more comprehensive and in-depth studies on vegetation, soil and livestock, exploring the mechanism of zoned rotational grazing from different perspectives and evaluating its advantages and disadvantages. Sweet studied the effect of grazing system on the degrada-

Chapter 1 Introduction

tion of Botawana grassland and found that the machine cover of rotational grazing was higher than that of free grazing at more than 2 km from the watering point, and the vegetation composition of rotational grazing was better than that of free grazing (Pulido & Leaver, 2003). According to Hart and his colleagues (1989), 62% of the total accumulation of hardy ribbon fern (*Pteris cretica L.* var. *nervosa*) production in mountainous areas under rotational grazing conditions was reported, while this value decreased to 23% under a continuous grazing system. Ralphs et al. (1984) showed that short duration rotational grazing can increase the yield of pasture and grassland utilization. Martin and his colleagues reported that rotational grazing can promote the recovery of grassland vegetation, increase grass coverage and pasture quality (Allen, 1985). The results of many studies show that zoned rotational grazing can improve grassland and prevent grassland degradation (Han et al., 2003; Wei et al., 2003). In particular, selecting suitable grazing systems according to different site conditions can significantly improve the efficiency of grassland utilization, which is beneficial to improving grassland and increasing livestock production. Other studies have also confirmed the superiority of zoned rotational grazing from different perspectives. Examples are the study of cattle feeding conditions under short rotation grazing in desert grasslands by Gammon and Roberts, the study of grassland improvement and livestock production under different grazing systems in the North American Prairie by Galt and James, the study on pasture and livestock production under different grazing systems by Allan, the study on the effects of fertilization and grazing systems on livestock production by Clark et al., the study on grazing systems, pasture area, livestock behavior, vegetation use status and livestock production by Hart et al., the study of biomass, mass and above-ground standing dead matter of available forage grasses under different grazing systems by Heilschmidt et al., and the study on the comparison of grazing regimes on mixed artificial grasslands by Hoveland

Evaluation Research on the Benefits of Grazing System in the Mongolian Grassland

(He et al., 1980). The above studies fully demonstrate that the zoned rotational grazing system has been widely accepted and applied in most countries.

In China, the research on zoned rotational grazing systems started late. The trial of zoned rotational grazing in Yanchi County of Ningxia Hui Autonomous Region and Guide County of Qinghai Province is a preliminary attempt to explore the rotational grazing system in China. In 1964, Ren and his colleague (1964) had systematically and comprehensively elaborated on the theory and method of zoned rotational grazing, and had carried out a series of zoned rotational grazing experiments using yaks and Tibetan sheep and other livestock, hence obtained a large amount of research data. Also during that period, some sporadic experiments were carried out in some other areas to explore the effects of zoned rotational grazing systems. Since the 1980s, due to the gradual accentuation of grass – livestock conflict and the increasing degradation of grassland, China's scholars have been increasing their research on grassland zoning and rotational grazing. The study conducted by Xu et al. on fenced grazing showed that fenced grazing can protect pastures and improve livestock production, hence suggesting that fencing is one of the important means to use grasslands in a rational and planned manner (Wang, 1996; Zhang, 1995; Chang, 1998). The results of a rotational grazing experiment conducted by Shi et al. (1983) in Sichuan Province showed that compared with free grazing, a rotational grazing system for cold season could improve grass production and stocking rate. Since the 1990s, as people become more aware of grassland conservation, corresponding research is gradually increasing. For instances, Li studied the long – term effects of different grazing regimes on replanted grasslands in southern New Zealand, and the results showed that rotational grazing promoted the growth of most grass species; high – intensity continuous grazing contributes to the invasion of weeds such as cotton weed *Pseudognaphalium affine* and narrowleaf hawkweed *Hieracium umbellatum*, accelerating the deg-

Chapter 1 Introduction

radation of grassland (Liu, 2006); Li (2010) studied zoned rotational grazing at the Sagebruch Desert in the Northern Slope of Tianshan Mountain; Liu (1993) and Geng & Ma (1993) studied on zoned rotational grazing in artificial gasslands; Wang (1995) discussed the design method and linear planning of zoned rotational grazing; Han et al. (1999) conducted an experimental study on zoned rotational grazing for beef cattle breeding; and Wang et al. (1999) studied the different grazing systems in typical grasslands of Inner Mongolia.

Grazing systems can also be classified as nomadic grazing, semi-nomadic grazing, and sedentary grazing, depending on whether or not human habitation is fixed. The grasslands of the Mongolian plateau were one of the grasslands that were used for aspects very early in history, where there were generations of nomadic pastoralists. Nomadic pastoralism, an ancient and traditional way of production and life formed in grassland areas, was naturally formed in the process of livestock domestication by human beings, and a grazing system was established after a series of evolution and development. Nomadic grazing is the most primitive form of grass grazing system and is part of the evolutionary process of recognizing interdependence and coevolution in the natural world. Thus, the nomadic grazing system is actually a primitive system of rotational grazing of grasslands. The nomadic system has many advantages because nomadic herds do not graze in the same pasture for long periods of time, hence avoiding degradation of pastures. Semi-nomadic grazing system refers to the grazing system in which a part of the population is nomadic and the other part is engaged in other work in a relatively fixed settlement, and it is a transformation from nomadic to settlement. In the early days of the People's Republic of China, pastoralists in Inner Mongolia lived a traditional nomadic life, but later, due to the need for institutional management such as household registration, their lifestyle gradually changed from nomadic to semi-nomadic and settled. Under a sedentary grazing system, pastoralists usually have a fixed residence, a fixed group of livestock, and even a fixed range of pastures. Sedentary grazing system includes continuous grazing and zoned rotational grazing.

1.2.3 The Study on the Effect of Grazing on Grassland Communities

The vegetation of grazing pastures is usually grasses, but may also include some other families of plants, such as Cyperaceae, growing in large grazing areas. Artemisia is the dominant plant in many pastures with sufficient moisture and high grazing pressure, especially in alpine meadows. Herbaceous and shrubs which are halophyte (salt - tolerant plant), especially Chenopodiaceae, are widespread on alkaline and saline soils in arid and semi - arid grasslands. In tundra areas, lichens, especially *Cladonia* spp. and bryophytes, are the main food source for reindeer. In addition, subshrubs are also very important, and the distribution of different species of Artemisia ranges from North Africa eastward to the northern line of the Steppe and North America. The subshrubs of the family Cuculidae (*Calluna* spp.), *Erica* spp., *Vaccinium* spp., etc, are the main food sources of goats and deer in the wetlands of England. The shrubs are often used as an important source of fodder and are generally eaten by livestock during the lean season and, in some cases, their fruits can be eaten too. Woody fodder is particularly important during the alternating wet and dry seasons in the tropics and subtropics. Different mixed shrub systems (Mediterranean evergreen shrub, Mediterranean coastal shrublands) are used for grazing in the Mediterranean zone. Trees and shrubs, especially willows (*Salix* spp.), are also important winter fodder in some cold regions.

Large areas of grassland are used for a variety of purposes in addition to being an important source of fodder for livestock and a major source of livelihood for farmers and pastoralists. Most grasslands are important water areas and the management of their vegetation is particularly important for the water resources of lands downstream. Poor grassland management can destroy grasslands, and the destruction of grasslands can increase soil erosion and runoff, severely damage agricultural land and its infrastructure, and cause siltation of irrigation systems and reservoirs. Good water catchment management is also beneficial to people living outside the grassland area, but it needs to be maintained by the farmers and pastoralists living in this grassland. These grasslands are also a major genetic source of biodiversity, not only providing important wildlife habitat, but also facilitating the in situ conservation of these genetic resources. In some regions, grasslands are important tourist and recreational sites, as well as important sites for religious activities. Among other fields, the grasslands are also a source of wild plants, natural medicines and other products.

Grasslands are a huge carbon reservoir at the global level. Minahi et al. (1993) argue that grasslands are almost as important as forests in terms of greenhouse gas cycling, and that underground organic matter in grasslands is as important as trees, which can increase the carbon storage capacity of grasslands.

Scholars around the world have conducted numerous studies on the effects of grazing on the quantitative characteristics of plant communities (Dong et al., 2011). More researchers believe that appropriate grazing intensity can increase the richness and complexity of community resources, maintain the stability of grassland plant community structure, and improve community productivity (Wang et al., 2001). Overgrazing can lead to degradation of grassland habitats, changes in community species composition, reduced diversity, and decreased productivity (Sa et al., 2010). There is much evidence that moderate grazing facilitates the maintenance of species vulnerable to human activities, thereby increasing local biodiversity; however, overgrazing (removing about 90% of above-ground biomass) will severely reduce the diversity of grassland plant species. Spence et al. (1971) argue that continuous heavy grazing can lead to significant decreases in vegetation coverage, height, standing stock and below – ground biomass. The study of Bisigato et al. showed that heavy grazing was more likely to change the plant plate structure than light grazing (BISIGATO) (Alejandro et al., 2005). The study of Austrheim and Eriksson (2008) on the changing patterns of plant diversity in the Scandinavian mountain range showed that grazing is a key process for maintaining biodiversity in the Scandinavian Mountain range. McIntyre and Lavorel (2001) showed that grazing changes the species composition, species richness, vertical structure, plant characteristics, and other attributes of grasslands. Altestor et al. (2005) argued that grazed grasslands have a higher species richness and species diversity than forbidden grazing grasslands, and grazing has led to the replacement of some shrubs species by creeping grasses.

At the end of the 18th century, following James Anderson' theory of rotational grazing, many people conducted experiments in different regions and supported his view that rotational grazing could improve pasture yield and grass utilization (Derner et al., 1994; Jacobo et al., 2006). It has also been shown that rotational grazing can promote the recovery of grassland vegetation and improve grass coverage and forage quality (Parsons, 1980). In particular, choosing suitable grazing systems according to different geographical conditions can improve the efficiency of grassland utilization, prevent grassland degradation, and benefit livestock production (Hou & Yang, 2006). However, Derek W. Bailey suggested that in arid and semi – arid areas, timely adjustment of grazing intensity is preferable to rotational grazing and grazing bans in maintaining and improving ecological health at regional and landscape scales (Bailey & Brown, 2011). The study of Martin et al. showed that rotational grazing can promote the recovery of grassland vegetation when the grassland condition is poor, but this effect is not signifi-

Chapter 1 Introduction

cant when the grassland condition is good. Heitschmidt et al. (1987) chose cattle as experimental animals for his study in the State of Texas and came to the following conclusion: The effects of rotational and continuous grazing on the vegetation environment were essentially the same, and the differences were mainly caused by differences in grazing intensity. In addition, there were differences in the effects of grazing seasons and grazing systems on the vegetation produced.

1.2.4 The Study on the Effect of Grazing Systems on Grassland Communities in Mongolia

Grazing is the main method of grassland use in Mongolia, and therefore the effect of grazing on grassland plant communities has become one of the most central elements of grazing ecology research (Han et al., 2005; Conte & Tilt, 2014).

Many Chinese and foreign scholars have conducted numerous studies on the effects of grazing systems and vegetation communities in Mongolian grasslands. Wei et al. (2000) analyzed the dynamics of plant communities under different grazing systems in Mongolian desert grasslands. The results showed that the density, height, coverage and importance values of dominant species were higher in the zoned rotational grazing areas than in the free grazing areas, while the height, coverage and density of some degraded plants, annuals and broad-leaved herb increased in the continuous grazing areas. Yang and Ba (2001) conducted a comparative analysis of the reproductive characteristics of the main plant populations, the needlegrass, in desert grassland under zoned rotational grazing and continuous grazing conditions. The results showed that the zoned rotational grazing system was more favorable to the formation of reproductive branches of needlegrass and could produce more seeds than the continuous grazing system, and the zoned rotational grazing was favorable to the survival of the seedlings of the main population in the desert grassland. Li et al. (2004) studied the effects of two grazing systems, ie. the zoned rotational grazing system and the continuous grazing on desert grassland plant communities and found that the rotational grazing system had less effect on the grass community. Han et al. (2004) studied the effects of two different grazing systems, zoned rotational grazing and continuous grazing, on sheep feeding and body weight. The results showed that continuous grazing system is not conducive to the uniform use of pastureland, so that the grass nutrition and sheep weight show large fluctuations, while zoned rotational grazing can maintain a steady increase in sheep weight and higher levels of livestock production. Zhu (2001) conducted a comparative study on the effects of different grazing systems on the above – ground biomass of three major plant populations: *Stipa breviflora* (needlegrass), *Cleistogenes songorica* (a type of flowering grass), and *Allium polyrhizum* (Asian species of wild onion) in desert grassland communities. The results showed that forbidden grazing can increase the above – ground biomass of community populations, and rotational grazing is more conducive to the recovery and increase of above – ground biomass of community populations compared to continuous grazing. By studying the effect of grazing system on plant growth in desert grassland communities, it was concluded that the plant growth rate in rotational and forbidden grazing areas is greater than that in continuous grazing areas. Na and Myagmartseren (2018) conducted a comparative study on the community characteristics of large needlegrass grassland under different grazing systems and discovered that the community plant species richness and diversity of zoned rotational grazing under moderate grazing pressure are higher than that of free grazing, the number of population structure relationship within the community is more complex than that of free grazing, and the decline in community plant evenness is less than that of free grazing. Peng and Wang (2005) explored the effects of different grazing systems and fencing on degraded grassland vegetation and found out that under the same livestock rate, the zoned rotational grazing system can improve the frequency, coverage, importance value and biomass of forage grasses compared with continuous grazing through the reasonable allocation of grazing time and space. Besides, its ecological restoration effect on grassland is the same as that of fenced grazing, hence zone rotational grazing is relatively the optimal grazing system. Osterheld and McNaughton (1991) studied the effects of different grazing systems on the population dynamics and plant compensatory growth in a family's meadow grassland, and the results showed that zoned rotational grazing was more effective than free grazing in increasing the amount of existing stock, growth and productivity. The study of Wu et al. (2005) on the effect of different grazing practices on vegetation characteristics in Ningxia arid grasslands showed that the implementation of zoned rotational grazing in six zones was the best way to use this type of grassland scientifically, and the productivity of rotational grazing was increased compared to the control grassland.

In summary, in recent years, precision grazing management has become the focus of grazing ecology research in China and abroad, among which, rotational grazing and continuous grazing are the hot research topics, which mainly using grazing control experiments to explore the advantages and disadvantages of rotational grazing and continuous grazing. Although the grazing control experiment is highly operational, there are still the following problems: ① basically choose the growing season, rarely consider four – seasons rotational grazing; ② more studies that have been conducted only selected one type of livestock, without considering the combined effects of several livestock; ③ small spatial extent and short time span; ④ simple control of grazing and leisure time without considering the effective regulation of grazing production by grazing operators according to climatic conditions and pasture conditions.

Mongolia and China's Inner Mongolia Autonomous Region together occupy the main part of the Mongolian steppe. Given the overall physical geography of the Mongolian grasslands and the consistency of previous grazing systems, as well as the subsequent implementation of different grazing systems in the two countries (seasonal rotational grazing in Mongolia and sedentary continuous grazing in Inner Mongolia Autonomous Region of China) that have produced large differences in the same grassland plant communities, the China – Mongolian cross – border region is an ideal site for the study of differences in grassland ecosystem responses to different grazing disturbances (Sa et al., 2010). However, due to various reasons such as the vast territory, national boundaries, communication barriers and tough environment, there are still lack of in – depth research on Mongolian Plateau's resources and environment, especially the systematic and comprehensive scientific investigation, the accumulation of materials, the spatial distribution pattern at different scales, the differentiation pattern and the in – depth driving and response research. Besides, most studies have focused on the Inner Mongolia region of China, and relatively few studies have been conducted on the territory of Mongolia. Those researches focused on atmospheric circulation, historical climate change, sandstorm and the distribution of some specific plant species, plant physiological and micro – geomorphology, species evolutionary branching systems, etc. In addition, there are some studies on the distribution patterns of vegetation growth and health conditions on the plateau, the long – time series responses of vegetation parameters to climate change, the measurement of local soil and wind erosion rates, and the assessment of regional wind erosion risk (Liu et al., 2007). While botanists' analyses usually lack a picture of the overall divergence patterns on the Mongolian Plateau, geographers' large – scale analyses are slightly less supported by primary data. On the other hand, most current studies on the characteristics of grassland vegetation, due to the hierarchical nature and complexity of ecosystems, researchers do not have a uniform understanding of the relationship between the number of grassland plant species, biomass and environmental factors in different regions and at different scales (Ba et al., 2015). The typical steppe is the most widely distributed and representative type of steppe in Mongolia. Its ecological uniqueness, specific species composition, community type, structure and function also show the fragility of this ecosystem. Because of the influence of natural and human factors, the degree and rate of degradation of typical steppe, which is mainly used for grazing, is increasing. Therefore, the study of typical steppe vegetation is conducive to the conservation of grassland resources and the maintenance of stable development of livestock industry in Mongolian grasslands.

Studies conducted in Inner Mongolia have focused on plant community change processes under different grazing intensity (Xiao et al., 2013; Cong et al., 2017). However, there are few comprehensive and in–depth studies on the process of different grazing practices on plant community diversity in typical steppe, especially studies on cross–border areas of Mongolian grasslands. Most studies on different grazing systems focused on analyzing the effects of grazing disturbance on vegetation communities through controlled experiments at the site scale (Li, 1995; Zhang, 2006; Yang et al., 2015). Therefore, there is a lack of research on the differences of different grazing systems on plant communities under natural conditions, especially the lack of multi–scale analysis such as quadrat, belt transect, community and region, as well as quantitative analysis such as remote sensing of earth surface and natural–social integration.

Future research should be conducted based on the law of grassland ecosystem development, with the goal of pursuing a balance between ecological protection and economic development, and minimizing grassland degradation and system destruction. Therefore, it is urgent to carry out basic science and applied research in line with the characteristics of the grassland geographical environment.

1.3 Research Background and Significance

1.3.1 Research Background

In the past 30 years since the implementation of continuous grazing system in Inner Mongolia Autonomous Region, the impact of grazing and global changes on the degradation of Mongolian grasslands has become increasingly serious. It resulted in serious disturbance and destruction of the balance and stability of the grassland ecosystem, which in turn has significantly reduced the service value of grasslands, shrunk the livestock economy, gave rise to many social problems and even formed a poverty trap. Therefore, how to break this predicament has become the primary problem that governments at all levels and relevant scholars need to solve and consider.

At present, there are many views in the academic and social circles about the causes of rapid deterioration of the Mongolian grasslands ecological environment, especially the Inner Mongolia grasslands. Among which, in addition to the global climate change as a cause, the mainstream view is overgrazing, thus proposing the implementation of grassland law controlling the balance between the grass yield and the number of livestock raised and emigration of population in order to achieve a harmonious relationship between human and pastoral land issues.

Related studies show that most theoretical livestock stocking rate of grasslands in Inner Mongolia Autonomous Region dropped by half in the 21st century compared to the 1960s, while the actual livestock stocking rate exceeded the theoretical capacity in the late 1990s. In other words, the theoretical livestock stocking rate of Inner Mongolia's pastoral grasslands has been declining for more than 30 years without overgrazing. What are the reasons for this phenomenon? Obviously, important causes other than overgrazing may also contribute to grassland degradation. Besides, what is the main reason that causes rapid deterioration of the Mongolian grasslands ecological environment? What is the deterioration mechanism? How to achieve a harmonious relationship between the human population in Inner Mongolia and nature? These are the complex process of interconnection, control and interaction of natural and human factors, which is a scientific issue with important theoretical and practical significance. The authors attempt to reveal the causes of the rapid deterioration of the Mongolian steppe ecology and its mechanisms through a comparative study of the China – Mongolian border, and to propose scientific countermeasures and recommendations for regulation.

1.3.2 Research Significance

In the empirical study of such a special ecosystem like Mongolian pastoralism, which is extremely fragile but has a history of thousands of years of harmonious development of man and nature, yet has been severely damaged in modern times, it is possible to obtain ground – breaking theoretical results from a comparative study of the grassland environment under different grazing systems in Mongolia and Inner Mongolia Autonomous Region. The grassland ecological barrier of the entire Mongolian plateau, including Inner Mongolia Autonomous Region, has been an important ecological guarantee for the survival and development of Asian peoples for thousands of years, and is also an important ecological barrier for future survival and development. Harmonizing the relationship between people and land of the Mongolian plateau, re – establishing a benign mechanism for the operation of the human – land system on the Mongolian plateau, and achieving sustainable development of Mongolian pastoral areas are the fundamental interests of the ecological security of the entire planet. Therefore, it is of great theoretical and practical importance to study the differential effects of different grazing systems on grasslands in Mongolian pastoral areas.

1.4 Research Content and Purpose

1.4.1 Research Content

Firstly, the study on the diversity of vegetation communities. A comparative analysis of changes in vegetation community diversity under different grazing systems in typical grassland areas of the Mongolian plateau using the diversity index method to explore the main causes of changes in plant diversity, grassland productivity and grassland condition.

Secondly, the study on the dominance of the main pioneer plants species. A determination analysis was conducted to identify the dominance of major pioneer plant species, and a statistical analysis to compare the effects of different grazing systems on the dominance of major plant species (the pioneer species), dominant species and annuals and broad – leaved herb in typical steppes of China and Mongolia.

Thirdly, the study on the characterization of the functional groups of the plants on the Mongolian plateau. Changes in functional group diversity can more effectively reflect changes in grassland vegetation. A comparison of typical grassland vegetation communities in China and Mongolia from the perspective of functional groups provides a basis for determining the trends of vegetation communities under different grazing systems.

Fourthly, the evaluation of the stability of typical grassland ecosystems on the Mongolian Plateau. Based on the analysis of vegetation community coverage, dominance and diversity, the stability of typical grassland ecosystems under different grazing practices in China and Mongolia was evaluated by using the M. Godron's community stability test method.

Fifthly, the calculation of the time series vegetation NDVI of the study area, analysis of vegetation dynamics under different grazing systems in China, Mongolia and the border area, to provide a basis for analyzing the causes of dynamic changes in grassland ecosystems.

1.4.2 Research Purpose

The study in this book aims to elucidate the evolution of grazing systems and differences in grazing ecosystems on the Mongolian plateau; to elucidate the differences in productivity and grassland stability of typical grassland meadows under different grazing practices on the Mongolian plateau; to elucidate the differences in biodiversity and functional plant characteristics of typical grassland vegetation under different grazing systems on the Mongolian Plateau; to elucidate the typical grassland vegetation succession trends under different grazing practices on the Mongolian Plateau; to exploring the causes of vegetation degradation and analyzing the effects of different grazing systems in China and Mongolia on the grassland ecosystems of the Mongolian plateau; and to promote the integration of grassland ecology, geography, economic management, remote sensing and spatial science to promote multidisciplinary research on grassland environment.

Chapter 2　Study Area

2.1　An Overview of Mongolian Plateau

2.1.1　Natural Condition

(1) Geographical Location. The Mongolian plateau is located at the Central Asian Plateau with a latitude and longitude range of 37°24′ ~ 53°23′N; 88°43′ ~ 126°04′E. The plateau is enormous with an area of about 2 million square kilometers as it extends latitudinally from Daxing' anling Mountain range in the east to the Altai Mountain range in the west, and extends longitudinally from Sayan mountain range and Yablonoi mountain range in the north to the Yin mountain range in the south. The plateau is politically including the whole country of Mongolia; Tuva Republic and Republic of Buryatia in the southern part of Russia; and the entire territory of Inner Mongolia Autonomous Region at the northwest and parts of Xinjiang Uygur Autonomous Region in China. This study se-

lected the core of the Mongolian plateau located in Mongolia and Inner Mongolia Autonomous Region as the study area.

(2) Topography. Some studies suggest that the uplift of the Mongolian Plateau is related to the collision of the Eurasian and Indian plates and is a by-product of the uplift of the Qinghai-Tibet Plateau, with its main uplift time concentrated in the last 10 million years (Wen, 2015).

Most parts of Mongolian plateau are ancient platforms with an average altitude of 1,580 meters, and its topography altitude decreases gradually from west to east. The eastern edge of the Mongolian Plateau is the Daxinganling Mountain range, which stretches 1,400 kilometers from northeast to southwest. The southern edge of the Mongolian Plateau is the Yin Mountain range and the Helan Mountain range. The Yin Mountain range is a complex folded fault mountain range that stretches for more than a thousand kilometers from east to west. The Yin mountain range which consists of the main Daqing Mountain, hills and low mountains, is the watershed between the Pacific and inland basins of the Mongolian Plateau, and is the natural boundary between the natural landscape of the Mongolian Plateau and the northern forest areas of the country. The eastern edge of Alashan Plateau is Helan Mountain range and its southern edge has Qilian Mountain range, northern mountains and Hexi Corrido distributed. The western edge of the Mongolian Plateau is the Mongolian Altai Mountain range extending northwest-southeast. With an average altitude of 3,000-3,500 meters, it is the highest mountain range on the Mongolian Plateau, which has become the border of the Irtysh Valley.

The Mongolian Altai Mountain range extend eastward to the Gobi Altai Mountain range in the central region of the Mongolian Plateau, which consist of the parallel Mongolian Khentii mountain range and the Khangai mountain range, which form the watershed between the Arctic Ocean basin and the inland basin of the Asian continent. The northern part of Khangai mountain range in Mongolia, are groups of low mountains and the highland on its north are connected with the eastern Sayan mountain range and the Tannu – Ola mountain range, in the middle of which a sunken basin is formed with Lake Khuvsgul as the center.

(3) Climate. The Mongolian plateau is influenced by the Siberian – Mongolian high pressure in winter, which makes it dry and cold, and it is also a necessary route for the latitudinal circulation, and the westerlies are extremely active. In summer, it was hot and rainy due to the East Asian monsoon (Dan, 2014). The precipitation on the Mongolian Plateau is much lesser than the evapotranspiration, and most of its area is arid and semi – arid. With a temperate continental climate, it is cold in winter and hot in summer. Precipitation mainly occurs from May to October, and its precipitation accounts for more than 90% of the annual precipitation. Because the eastern and northern parts of the plateau are influenced by water vapor from the Pacific and Arctic Oceans, precipitation decreases from east to west and from north to south. Meanwhile, the temperature shows the opposite change characteristic, ie. The temperature gradually decreased from the southwestern part of the plateau to the northeastern part of the plateau. The distribution of the total solar radiation shows the same characteristics as the temperature, with the total annual radiation in the southern Alashan region being about 160 Kcal/cm^2, while the total annual radiation in the northeastern and northern mountainous regions is 120 – 130 Kcal/cm^2. The humidity in the northern part of the Daxinganling and Khangai and Khentii mountain range is 0.5 – 0.8, in the Alashan region it is 0.03 – 0.13, and in the rest of the intermediate areas it is 0.13 – 0.5. Precipitation on the Mongolian Plateau varies greatly from year to year and is unevenly distributed within the region. The average annual temperature is – 6.6℃ – 3.9℃, extreme high temperature is 40℃, extreme low temperature is – 52℃, absolute temperature difference can be up

to 90℃, and the daily temperature difference can be up to 20℃ – 30℃. The climate of the small region changes with the change of altitude and the distribution of mountains. Most days of the year are sunny, the atmosphere is dry and precipitation is scarce, and the seasonal changes of the Mongolian plateau are very obvious, with cold and long winters, windy springs, hot summers, cool autumns, and relatively humid summers and autumns.

The climate type of Mongolian Plateau changed from humid and semi – humid climate in the southeast to semi – humid and semi – arid climate in the central and to semi – arid and arid climate in the northwest (Qin, 2019).

Precipitation moisture in the eastern and southeastern part of the Mongolian Plateau (Inner Mongolia) comes from the Pacific Ocean and decreases from 400 – 600 mm in the northeast, 300 – 400 mm in the south and southeast, to 100 – 200 mm in the north and northwest; at the border between China and Mongolia, precipitation is around 200 mm. The rain water of the northern Mongolian plateau (Mongolia) is mainly sourced from the Arctic Ocean and the precipitation decreases from 300 – 400 mm in the north to about 100 – 200 mm in the south.

(4) Hydrology. The Mongolian Plateau water system is mainly divided into the Arctic Ocean Basin, the Pacific Ocean Basin and the Central Asian Inland Basin. The Khangai and Khentii mountain range are the source of the rivers of the Arctic Ocean; the rivers of the Pacific basin also flow through the Khenthii mountain range, the eastern plains of the Mongolian Plateau, and the spur of the Daxinganling mountain range and the piedmont region; the Central Asian Inland Basin includes parts of the Khentii mountain range and the Khangai mountain range, the Gobi Altai mountain range, the Gobi Desert, and the Mongolian Altai mountain range. The high permeability of the soil promotes the formation of the piedmont region and the water table of mountain on the Mon-

golian plateau, and therefore the rivers in the region are characterized by relatively high water table supply.

(5) Soil. Chestnut soil and chernozem are the main soil types on the Mongolian plateau while brown pedocals are developed in arid areas, and gray – brown desert soils are developed in extremely arid areas. Daxinganling and Yin mountain ranges are mainly distributed with brown earths and grey forest soils, and black soils are widely developed in the hilly areas in front of the mountains.

Chestnut soils are widely distributed in the eastern plains and southeastern regions of the Mongolian Plateau. The lower part of the broad valley and slope of the mound is dominated by dark chestnut soils, and meadow soils on the river bank. The southeastern part of the steppe zone consists of aeolian soils and sandy meadow soils. Central Lake Valley, East Gobi and Ulaanbaatar Plateau are widely distributed with brown pedocals. Grey – brown desert soils, grey desert soil and aeolian soils are widely distributed in the Altai Gobi and Alashan regions.

(6) Vegetation. Different climatic zones foster different ecosystems and vegetation type zones. The broad semi – arid climate zone forms the desert – steppe vegetation type, the hyper – arid climate zone forms the desert vegetation type, and the semi – humid climate zone forms the forest, meadow – steppe and grassland vegetation. The mountain forests are mainly composed of spruce and larch forests, and there are also secondary birch forests in the Daxinganling mountain range in the northeast and the Yin mountain range in the south. Meadow – steppe vegetation is widely developed in piedmont valleys and hilly areas. The vast plains in the eastern and southeastern parts of the Mongolian Plateau are continuously covered with large areas of feathergrass grassland (*Stipa grandis*), Chinese rye grass grassland (*Leymus chinensis*) and needlegrass grassland (*Stipa sareptana*). *Stipa baicalensis* grassland and *Filifolium sibiricum* grass-

land on gravel slopes while typical meadow and saline meadow vegetation are distributed on the river mudflats.

The sandy land has various meadow vegetation, and sandy vegetation widely distributed. Examples are elm sparse forest, spruce forest, scotch pine forest, *Caragana* scrub, meadow, *Artemisia* spp., marsh, and willow scrub. Influenced by climate, especially by precipitation, the vegetation cover spans from north to south in the order of forest, forest – steppe, typical steppe, desert – steppe, Gobi – desert, typical steppe and agro – pastoral interspersed area (unnatural pattern).

(7) Natural Resources. The Mongolian plateau is rich in natural resources, in addition to vast grasslands, deserts and Gobi, there are also forests and many kinds of mineral resources. In 2016, Mongolia had 150,000 hectares of forests, accounting for 9.6% of the country's land area, of which Siberian larch accounted for about 73%, cedar for 11% and other trees for 6.5%, with overall timber reserves of about 13 billion cubic meters. Mongolia's rivers and lakes have more than 50 native species of fish. The most abundant resource of the Mongolian plateau is minerals, which cover 465,000 square kilometers, or 30% of the territory. In 2016, registered mineral resources of Mongolia are 1170 deposits of 80 minerals, including coal, fluorite, copper, iron ore, gold, aluminum, lead, oil, tin, phosphate, uranium, tungsten, etc.

The Inner Mongolia Autonomous Region is a vast region, extremely rich in coal resources. The region's cumulative proven reserves have been the first in the country, as of 2016, Inner Mongolia Autonomous Region surveyed a total of 103 coal – bearing basins, cumulative proven reserves of 808.65 billion tons. The Inner Mongolia Autonomous Region produces and supplies 1/4 of the country's coal resources, making a great contribution to national social and economic development. Inner Mongolia Autonomous Region coal resources have four advantages: First, most of the coal fields have large reserves, thick coal seams, shallow burial, simple structure, stable occurrence, suitable for open – pit mining and large – scale mechanized mining, hence, coal mining costs are the lowest in China. Second, the coal quality and type are excellent. There are high quality anthracite coal, the Taixi Coal of Alashan League with low sulfur, phosphorus, ash and high heat generation, and high quality non – stick coal, the Ulan Coal of Ordos with extra low sulfur, phosphorus, and ash, and other coal types such as lignite are also good coal for power and chemical industry. Again, industry extension advantage. Coal products can become an industrial raw material with high added value through chemical or physical methods. High efficiency flue gas desulfurization, wet scrubbing, coal liquefaction, coal to oil and other technologies are becoming increasingly mature, so that the coal – based industry has broad prospects for development. Although the coal chemical industry in Inner Mongolia is still in its initial stage, the coal type, resource advantages and geographical area make it occupy a pivotal position in China, and it can be said that the development of its coal chemical industry has been a matter of the overall development of the national coal chemical industry. Finally, the advantages of human resources. After decades of development and construction, Inner Mongolia coal industry has a team of excellent professionals from coal exploration, design, construction, production to scientific research, hence, having the advantage of human resources to develop and build the coal industry. At present and in a certain period of time in the future, coal as the main energy source occupies an irreplaceable and important position in the energy structure of China.

2.1.2 An Overview of Socio – Economic

(1) An Overview of Mongolia. With an area of 1, 566, 500 square kilometers, Mongolia is the second largest landlocked country in the world after Kazakhstan and the

Chapter 2 Study Area

19th largest country in the world in terms of land area.

Most of Mongolia's land is covered by grassland, with little arable land. Gobi Desert is located at the south, and mountain ranges in the north and west. The pillar of Mongolia's national economy is grassland-based animal husbandry, with about 40% of the population engaged in pastoralism. In the past, Mongolia was an agricultural cooperative under a macro-planned economy, and the use and protection of natural pastures were carried out under a certain degree of planning, including long-distance migration, relocation, and loan of pastures. After the transition to a market economy, Mongolia has not made major changes in the way it uses and protects its natural pastures. Mining, agriculture and livestock, transportation and services are the main industries in Mongolia and the transition to a market economy started in 1991. The government formulated and implemented the 'Program for Privatization of State Assets 1997 – 2000' in July 1997 in order to make private economic components dominant in the national economy. As shown in Figure 2 – 1, Mongolia's gross domestic product (GDP) has largely maintained a continuous upward trend over the past 20 years, with a GDP of 2,394.28 billion tögrög (about RMB 73.7 billion, at the average exchange rate in 2016) in 2016, 19 times higher than in 2000. The GDP growth rate during the period was the smallest in 2009, 35 billion tögrög more than in 2008. As shown in Figure 2 – 1, the trends in Mongolia's GNI and GNI per capita are generally consistent with GDP, with the smallest growth rate of 0.5% in 2009 compared to 2008 and the fastest growth rate of 35% in 2011 compared to 2010. Gross national income increased from 1,218.5 billion tögrög in 2000 to 219.71 billion tögrög in 2017 (approximately RMB 676 billion at the exchange rate of that year); gross national income per capita increased from 509.7 thousand tögrög in 2000 to 7,113 thousand tögrög in 2016 (approximately RMB 21,918), an increase of 19.7 times in 20 years.

· 43 ·

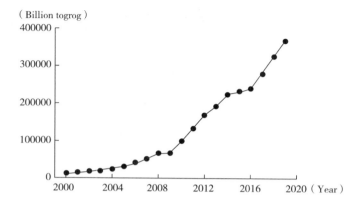

Figure 2 – 1　Gross domestic product of Mongolia

Source: http://www.nso.mn/.

As shown in Table 2 – 1, Mongolia's population was 648,000 in 1918 and grew to 3,297,000 in 2019, with 1,677,000 women and 1,620,000 men. It was an increase of 2,649,000 people from 1918 to 2019, a net increase of about 26,000 people per year. Small population size and sparsely populated land are important factors limiting Mongolia's socio – economic development.

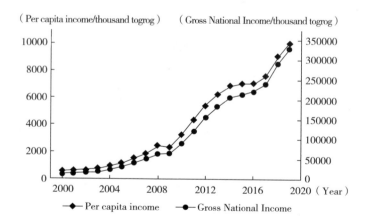

Figure 2 – 2　Trends in Gross National Income and Gross National Income per capita in Mongolia

Table 2-1 Total population change in Mongolia from 1918 to 2019

Unit: 10,000 people

Year	1918	1935	1944	1956	1963	1969	1979	1989	2000	2010	2019
Total population	64.8	73.88	75.98	84.6	101.7	119.8	159.5	204.4	237.4	275.5	329.7

Source: http://www.nso.mn/.

As shown in Figure 2-3, the proportion of urban population to total population in Mongolia has steadily increased over the last 20 years, but the proportion of rural population to total population has generally shown a decreasing trend. The number of rural population decreased from 1,042,000 in 2000 to 1,038,000 in 2019, and the urban population increased from 1,361,000 in 2000 to 2,259,000 in 2019. In 2016, the rural population was about 988,000 and the urban population was 2,131,000. With the urbanization of the population and the increase in the natural population growth rate increasing the pressure on the urban population.

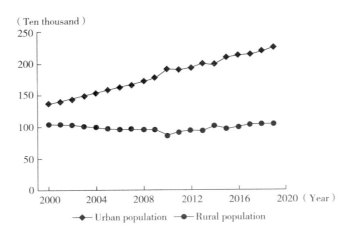

Figure 2-3 Trends in urban and rural population in Mongolia, 2000 to 2019

Source: http://www.nso.mn/.

蒙古草原放牧制度效益评价研究
Evaluation Research on the Benefits of Grazing System in the Mongolian Grassland

As shown in Table 2 – 2, the total number of livestock in Mongolia from 1970 to 2018 showed a general upward trend, reaching 66,460,200 heads in 2018, of which the proportions of sheep, goats, cattle, horses and camels were 46%, 40.8%, 6.6%, 5.9% and 0.7%, respectively. The total number of livestock increased by 45,971,000 head compared to 20,489,000 head in 1970, with an average annual growth rate of 224%. Although the total number of livestock in Mongolia is increasing, the Mongolian plateau has the saying of' there is no year without disaster and there are many disasters in a year'. Extreme weather and meteorological disasters are frequent, bringing huge losses to the development of the local livestock industry. According to the International Monetary Fund, 8.2 million head of livestock were lost due to drought and white plagues from 1999 to 2003 (Tachiiri et al., 2008), and the livestock mortality rate was as high as 23.4% in 2010 (Du et al., 2017). Both in the Inner Mongolia Autonomous Region of China and Mongolia, droughts and snowstorms have had a great impact on the production of Mongolian highland pastoralists and have become an important factor affecting socio – economic development.

Mongolia is located in the middle of Asia, with an average altitude of about 1,600 meters, and is a typical landlocked highland country. Agriculture and livestock are the basis of Mongolia's economy, with 17.6% of Mongolia's gross national product (GDP) in 2016 coming from agriculture and livestock, and 82.7% of the output of agriculture and livestock coming from livestock. Mongolia had only 508,000 hectares of arable land in 2016 (Figure 2 – 4), accounting for only about 0.3% of the country's land area. The arable land is mainly located in the Selenga, Tuv, Dordon and Bulgan provinces, where the Selenga River and its tributaries, the Orhon and Herlen Rivers, are located in relatively good water conditions, which together account for about 82% of the country's arable land. In the southern desert and Gobi regions such as Dornogovi,

Dundgovi and Omnogovi provinces, and in the western mountainous regions such as Bayan – Olgiy and Zavhan provinces, the arable land is less than 1% of the total arable land. Cereal crops are selected mainly based on their cold and drought – resistant ability, ie. wheat but there are also a small amount of barley and oats planted. Potatoes and

Table 2 – 2 Changes in the number of domestic animals in Mongolia from 1970 to 2018

Year	Horse	Cattle	Camel	Sheep	Goat
1970	231,900	2,107,800	633,500	13,311,700	4,204,000
1980	1,985,400	2,397,100	591,500	14,230,700	4,566,700
1998	3,059,100	3,725,800	356,500	14,694,200	11,061,900
1999	3,163,300	3,824,800	355,600	15,191,300	11,033,700
2000	2,660,800	3,097,600	322,900	13,876,300	10,269,800
2001	2,191,800	2,069,600	285,200	11,937,300	9,591,300
2002	1,988,900	1,884,300	253,000	10,636,600	9,134,800
2003	1,968,900	1,792,800	256,700	10,756,400	10,652,900
2004	2,005,300	1,841,600	256,600	11,686,400	12,238,000
2005	2,029,100	1,963,600	254,200	12,884,500	13,267,400
2006	2,114,800	2,167,900	253,500	14,815,100	15,451,700
2007	2,239,500	2,425,800	260,600	16,990,100	18,347,800
2008	2,186,900	2,503,400	266,400	18,362,300	19,969,400
2009	2,221,300	2,599,300	277,000	19,274,700	19,651,500
2010	1,920,300	2,176,000	269,600	14,480,400	13,883,200
2011	2,112,900	2,339,700	280,100	15,668,500	15,934,600
2012	2,330,428	2,584,621	305,835	18,141,359	17,558,672
2013	2,619,377	2,909,456	321,480	20,066,428	19,227,583
2014	2,995,754	3,413,851	349,299	23,214,783	22,008,896
2015	3,295,336	3,780,402	367,994	24,943,127	23,592,922
2016	3,635,489	4,080,936	401,347	27,856,603	25,574,861
2017	3,939,813	4,388,455	434,096	30,109,888	27,346,707
2018	3,940,092	4,380,879	459,702	30,554,804	27,124,703

Source: The website of National Statistics Office of Mongolia: http://www.nso.mn/.

vegetables account for about 3% of the arable land, and vegetable varieties are mainly turnips and carrots with a short growth period, followed by a small amount of onions and cucumbers.

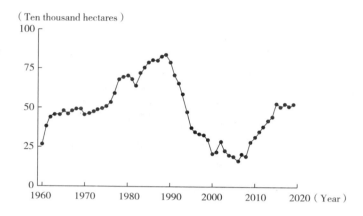

Figure 2 – 4   Change in arable land area from 1960 to 2019

Mongolia is rich in mineral resources, and some of its large mineral reserves are leading in the international arena. In 2016, Mongolia's industrial output reached 9,927.17 billion tögrög (30.59 billion RMB), of which the mining industry was worth 4,817.29 billion tögrög (about 14.84 billion RMB). At present, the main large and medium – sized mines in Mongolia that have been mined and exported are Oyu Tolgoi gold – copper mine (OT mine), Erdenet copper – molybdenum mine, Tavan Tolgoi coal mine (TT mine), Nariin – Suhai coal mine, Tumurjan Ovoo zinc mine, Baganuur coal mine, Tamchag oil field, etc.

(2) An Overview of Inner Mongolia Autonomous Region of China. It has been documented that 90% of the available natural grasslands in China have been degraded to varying degrees (Kawamura et al., 2005). The Inner Mongolia Autonomous Region cov-

Chapter 2 Study Area

ers an area of 1.183 million square kilometers, accounting for 12.3% of the total area of the country. The usable grassland area in Inner Mongolia Autonomous Region is 681,800 square meters, accounting for 57.6% of the total area of Inner Mongolia Autonomous Region, of which 75% of the grassland areas have experienced moderate degradation such as reduced vegetation coverage, desertification, salinization, etc (Xi, 2013). At the end of 2017, the resident population of Inner Mongolia Autonomous Region was 25,286,000, of which the urban population was 15,682,000 and the rural population was 9,604,000, and an urbanization rate of 62%. The male population was 13,052,000 and the female population was 12,234,000. In 1953, the Mongolian population was 683,000, accounting for 11.2% of the region's total population; in 2017, the Mongolian population was 4,639,000, rising to 18.4% of the region's total population, while the Han population fell from 87% to 74.4% of the region's total population. According to statistics, there are about 2.2 million Mongolians who speak Mongolian, which is roughly 50% of the total Mongolian population (Aruna, 2017). The total population of Inner Mongolia Autonomous Region grew by 19.186 million from 1953 to 2017, with an average annual growth rate of 295,000, or 315% per year. The changes in the total population of Inner Mongolia Autonomous Region from 1953 to 2017 are shown in Figure 2-5. The rapid growth of population in Inner Mongolia Autonomous Region is not only related to the natural growth of population, but more related to the several waves of reclamation since the founding of the People's Republic of China and also the construction of various projects and the resulting births population growth. For example, within 30 years after the founding of the People's Republic of China, planned migration to Inner Mongolia Autonomous Region became a major feature of the country's population policy, and Inner Mongolia Autonomous became the third most important population migration area in China. From 1954 to 1981, the net immigration to the region was about 3.14 million people, and this immigration from outside the province continued until 1981 when the Central Committee of the Communist Party of China made a decision not to migrate to Inner Mongolia Autonomous Region in large numbers (Wu, 2007).

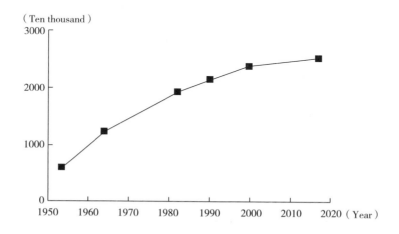

Figure 2-5　The changes in the total population of Inner Mongolia Autonomous Region

Source: *Information of censuses in Inner Mongolia Autonomous Region.*

As shown in Table 2-3, in 2019, Inner Mongolia achieved a regional GDP of 1,863.260 billion yuan, an increase of 7.2% over the previous year at comparable prices. Among them, the primary industry was 162.87 billion yuan, the secondary industry was 907.89 billion yuan, and the tertiary industry was 792.51 billion yuan. The ratio of the three industries was 8.8 : 48.7 : 42.5, and the contribution of the primary, secondary and tertiary sectors to GDP growth was 3.8%, 49.0% and 47.2%, respectively. In 2019, the per capita GDP of Inner Mongolia reached 74,069 yuan, an increase of 6.9% over the previous year; the annual per capita disposable income of all residents reached 24,127 yuan, an increase of 8.1% over the previous year, and an actual increase of 6.8% after deducting price factors; the per capita living consumption expenditure of all residents was 18,072 yuan.

Table 2-3 GDP of Inner Mongolia Autonomous Region

Year	1978	1995	2000	2005	2010	2018	2019
GDP	58.0	857.1	1,539.1	3,905.0	11,672.0	17,289.2	18,632.6
Primary industry	19.0	260.2	350.8	589.6	1,095.3	1,753.8	1,628.7
Secondary industry	26.4	308.8	582.6	1,773.2	6,367.7	6,807.3	9,078.9
Tertiary industry	12.7	288.1	605.7	1,542.3	4,209.0	8,728.1	7,925.1

Note: Data comes from Annual Statistic of Inner Mongolia (2019).

As shown in Figure 2-6, the total number of livestock in Inner Mongolia Autonomous Region from 1947 to 2018 generally showed a clear upward trend. In 2018, the total number of livestock in Inner Mongolia reached 72.779 million heads, including 7.787 million large livestock, 60.019 million sheep and goats, and 4.973 million pigs, accounting for 10.7%, 82.5%, and 6.8% of the total livestock, respectively. The total number of livestock in 2018 increased by 64,261,000 heads from 8,518,000 heads in 1947, an average annual growth rate of 754%. The livestock industry in Inner Mongolia Autonomous Region is greatly affected by extreme weather and natural disasters. For example, nearly 10 million heads of livestock died in Inner Mongolia Autonomous Region in 1982 due to drought, with a mortality rate of 24%. And the livestock mortality rate reached 38% in 1987 (Wu & Liu, 2004). Therefore, disaster prevention and mitigation becomes a top priority in the socio-economic development of the Mongolian plateau.

The change in arable land area is inseparable from population growth, and since the founding of the People's Republic of China, Inner Mongolia Autonomous Region's

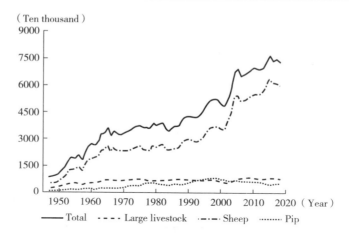

Figure 2-6 Changes in the number of livestock in Inner Mongolia Autonomous Region from 1947 to 2018

population and arable land area have shown rapid and steady growth (refer to Figure 2-7). In 2018, the arable land in Inner Mongolia Autonomous Region was 9.272 million hectares, 5.313 million hectares more than in 1947, 2.3 times more than in 1947, with an average annual growth rate of 99.7%.

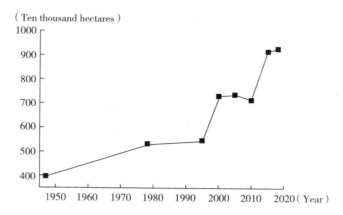

Figure 2-7 Changes in arable land area in Inner Mongolia Autonomous Region from 1947 to 2018

Source: *Inner Mongolia Statistical Yearbook* (2011, 2019).

Chapter 2 Study Area

2.2 The Evolution of Grazing System

2.2.1 The Evolution of Grazing System in Mongolia

The main production method on the Mongolian plateau is pastoralism, and the livestock include sheep, goats, cattle, horses and camels, which the Mongolians call the' five livestock.

Long – distance migration was a feature of early nomadic animal husbandry. At the beginning of the 4th century A. D. , the Rouran Khaganate migrated to the south of the desert in winter and retreated to the north of the desert in summer, with a nomadic radius of no less than several kilometres. At the end of the 4th century A. D. , Tuoba clan of Xianbei which established the Northern Wei, still crossed the Yellow River and moved their troops and livestock thousands of miles away from the present day Hetao region to station their herds (Hai, 2013).

About the mid 8th century, Genghis Khan's ancestor, Berdych and his wife, Wouei Malanle, migrated more than a thousand kilometres westward from the northern part of the present – day Daxinganling mountain range, along the present – day Ergune River, crossed Hulun Lake, and moved to the source of the Onon River for herding.

After the establishment of the Mongolian Empire in 1206, nomadic animal husbandry had made great progress. The grasslands had fixed seasonal camps whereby the pastures were used according to the seasons. Livestock would migrate to the mountains in spring and return to the plains in winter. In 1247, Zhang Yaoqing wrote a' Travelogue of the border beacons' that said the pastoralists are mostly on high and cold places in summer, and tend to move to a place with more forest cover and more sunlight to spend the winter (Zhang, 1936).

In 1275, the Italian traveller Mark Polo wrote in his travelogue that the Mongols did not settle in one place, but moved to a warmer plain at the beginning of winter in order to find a well-watered meadow for their livestock; in summer, they moved to a cooler part of the mountains where water and grass were plentiful and where they could avoid horse flies and other blood-sucking pests that would infest their livestock. They kept moving up the mountain for two or three months, looking for new pastures (Tang, 1981). Until the Qing Dynasty, due to the fixing of boundaries, the phenomenon of long-distance shift grazing no longer existed (Ge, 1996).

Han (1999) believes that the delineation of the boundaries of the pastoral areas of the leagues and banners has become a turning point for the ancient nomads in northern China from the nomadic lifestyle of following the water and grass and drifting back and forth to the way of living in a limited range. In the early Qing Dynasty to the Kangxi period, the Qing government changed the "ten thousand households system" from the Northern Yuan period to the "league-banner system". And the Mongolian region in the south of the desert was divided into six leagues and 49 banners, as well as the three other banners not under any league, Alashan Errut Banner, the Ejina Turks Banner and the Xilian Banner, a total of 52 banners. Each has a clear boundary between the banners, and strict legal provisions prohibiting pastoralists from grazing across the border. The Qing government's "league-banner system" had a profound impact on the pattern of human-land relations in Mongolian pastoral areas and the direction of their development. After the implementation of the league-banner system in Mongolia, the nomadic life of Mongolian pastoralists was strictly limited to the pastures under the jurisdiction of

Chapter 2 Study Area

each banner, and the traditional nomadic range was greatly reduced and fixed.

Since the nomadic range of each banner (tribe) was strictly regulated, the most fertile pastures of each banner were occupied by the princes and nobles of each banner. This change actually shrunk the actual nomadic distance of all pastoralists, including the princes and nobles, from hundreds to thousands of kilometres to only about a hundred kilometres. This fact of having political means to force change quickly changed the traditional livestock production and operation of the pastoralists, hence fully transforming the nomadic production method into a limited nomadic production method (History Compilation Committee, Department of Animal Husbandry of Inner Mongolia Autonomous Region, 2000).

Banner is the main administration of the Manchu government in Mongolia whereby the area of the banner is similar to the current county in Inner Mongolia Autonomous Region. Today in Mongolia, banners no longer exist because they have been merged into leagues, which are divided into soums. Soum is a much smaller unit, and there are more soums than the previous number of banners. The 18 provinces are divided into 225 districts (soums), which are in turn divided into production units. Each collective production unit has an administration office, boarding school, medical facilities, veterinary, communication, recreational facilities and stores (Timofeev et al., 2016).

Collective production units are divided into livestock production teams or groups, which are further divided into Suurig – individual units consisting of 1 to 4 families. The pastoralists systematically moving (or rest grazing) seasonally in the production unit area or, if necessary, in the soum area according to a reasonable route, and the large meadows separated from the grazing pastures are preserved and managed (Qing, 2014).

2.2.2 The Evolution of Grazing System in Inner Mongolia Autonomous Region

The two countries have adopted different grazing systems for grassland management: Mongolia has basically changed from nomadic grazing to semi-nomadic grazing or four-seasons rotational grazing, while Inner Mongolia of China has changed from nomadic or semi-nomadic grazing to sedentary continuous grazing. Especially after the implementation of the Pasture Household Contract Responsibility System in 1990, pastoralists began to fence their pasture land and graze their animals only at fixed locations (Na et al., 2018).

(1) Loss of quality pasture increases pressure on pastoralists to survive. Starting from the war and natural disasters in the late Ming and early Qing dynasties, a large number of bankrupt farmers moved from Hebei, Shanxi, Shaanxi, Shandong and Henan into pastoral areas, and reclaimed large areas of grassland, which accelerated the evolution of the structure of socio-economic and ecological systems in pastoral areas. In the past 300 years, the agricultural population in Inner Mongolia has grown rapidly, with the agricultural population of Chengde Prefecture alone, which was 550,000 in 1781, increased to 780,000 in 1827 and reached 2.82 million in 1901. According to statistics, from 1902 to 1928, 57 million mu (Chinese measurement whereby 1 mu equals to about 6 acres) of high quality grassland was reclaimed in Inner Mongolia today. In the process of losing pastureland, a significant portion of pastoralists have been completely transformed into farmers (Yun, 1999). The historical period when the total population of China first exceeded the 100 million benchmark was the Qing Dynasty, but the spatial distribution of the population at that time was very uneven. In 1749 (the 14th year of the Qianlong reign), when the population density in Shandong had reached 162.2 per-

Chapter 2 Study Area

sons/km^2, the population density in Fengtian province was 3.3 persons/km^2 (Hu & Zhang, 1985). In the corresponding period (1770), the average population density of Zhelimu League (now Tongliao City) was about 0.86 people/km^2 (Wang & Shen, 1997). The uneven spatial distribution of the population is a possibility for the natural movement of the population. In 1902, the Qing government launched the so-called "new policy" guided by the idea of "immigration to the real border". One of the main measures specifically implemented by the "new policy" in Inner Mongolia was to change the policy of "sealing off the Mongolian lands" to reclaim the Mongolian lands in order to increase the state's revenue through taxation with a view to paying the huge indemnities. This kicked off the full-scale reclamation of Mongolian land and triggered another wave of migration to Mongolian land. By the end of the Qing Dynasty, the total population of Zhelimu League had reached 2,493,000 people, of which the Mongolian population was only 193,000. The Mongolian population of Zhelimu League in 1770 (the 35th year of the Qianlong reign) was 183,000, which was about the entire population of Zhelimu League at that time. And the agricultural population crossing the border from mainland was already up to 2.3 million after more than 100 years. Hence, the average population density of Zhelimu League at the end of the Qing Dynasty had increased to 11.71 persons/km^2 (Wang & Shen, 1997).

With a large number of people poured into the Mongolian land, agriculture began in the Mongolian region and was developing rapidly. But up to this point, agriculture in pastoralist banners is only a supplement to the nomadic economy, and the traditional nomadic or four-seasons rotational grazing system is still predominant in the grassland pastoral areas.

(2) Increase in agricultural population, deepening into pastoral areas. At the beginning of the founding of new China, Inner Mongolia Autonomous Region became one

of the important population immigration areas. From 1954 to 1981, the net immigration of the whole region was about 3.14 million, and the immigration accounted for 28% of the net increase of 11.44 million people in Inner Mongolia Autonomous Region in the same period. These immigrants are widely distributed in Inner Mongolia Autonomous Region, especially in the southern and eastern parts of the region (Song, 1987).

In the first stage, population migration from other provinces continued until 1981 when the central government made a resolution to not migrate to Inner Mongolia Autonomous Region in large numbers, with 88% of the migrants moving in between 1954 and 1960 (Yu, 2006). The second stage was also dominated by the migration of large numbers of people from the interior provinces and supplemented by high rates of natural growth. Immigrants from agricultural areas of other provinces, together with a small number of previous immigrants, continued to move across the sandy grasslands of the south into the alluvial plains of the West Liao River basin in central Inner Mongolia Autonomous Region. In the third phase, the population mainly moves continuously from the southeast to the northwest, through a combination of immigration from outside the province, movement within the study area, and population reproduction. This migration was guided by the idea that "more people are more powerful" and "food is the key", which formed the solid basis for the later population growth and grassland degradation in Inner Mongolia Autonomous Region. Since the founding of the new China, the Inner Mongolia Autonomous Region has experienced (or is continuing to experience) the fourth peak of land reclamation. The previous three times were the period of economic recovery at the beginning of the founding of new China, the three years difficulty period, and the 10 years turmoil period, with a total of more than 2.5 million hectares of land cleared (Bao, 1999). Since the clearing of land was an organized movement under the relevant policies, it had a very wide impact. In Inner Mongolia Autonomous Region, there was a sharp increase in the area of arable land throughout the region. The change in the arable land area of the Zhelimu League can be seen in the decade or so from the mid – 1950s to the mid – 1960s (Yan & Xu, 2000), which was the first peak in the increase of arable land area in Horqin in the last 50 years, and the highest in the last 50 years. In 1960, the arable land of Zhelimu League had reached 63.5×10^4 hectares, an increase of 32% over the early years of the founding of new China, which accounted for about 11% of the total land area of the league. In Zhelimu League, except for the only West Liao River Plain, all of them are fixed and semi – fixed dune areas, and the reclamation during the peak period of land reclamation has advanced to the above – mentioned lots because the wasteland and pastureland that have been put into reclamation during the Republic of China period have been reclaimed. With the fourth wave of land reclamation that began in the late 1980s, the arable land area began to rebound again. In 1996, the arable land of the league reached 63.5×10^4 hectares, another 10% higher than the early years of the new China. In the grasslands, the bare land is highly susceptible to desertification. According to the study, the reclamation of one hectare of grassland in Inner Mongolia Autonomous Region can cause desertification of the surrounding three hectares of grassland. Therefore, the large – scale reclamation of grassland directly leads to the large – scale desertification of grassland, making the pasture for nomadic pastoralists smaller and smaller, and forcing them to change from nomadic grazing to continuous grazing or both farming and herding, or even completely engage in agricultural production (Wulan & Zhang, 2001). The proportion of Han Chinese in the pastoral population is rising rapidly. According to the population statistics of 33 pastoralist banners, the population increased by about 90,000 people per year during the three decades from 1950 to 1980. From 1950 to 1980, the proportion of Mongolians in Alashan decreased from 56% to 22% of the total population while the proportion of Mongolians

蒙古草原放牧制度效益评价研究
Evaluation Research on the Benefits of Grazing System in the Mongolian Grassland

in Xilingol League decreased from 90% to 28% of the total population. From 1949 to 1985, the grassland area of Inner Mongolia Autonomous Region decreased by 138 million mu, and although there were still pasture in the northern part for four-seasons rotational grazing, the nomadic range was only a few dozen kilometers in circumference, while the southern part of the pastoral area could not be nomadic at all (Ao & Da, 2003).

In short, due to the long-term combined effect of the change of traditional livestock production methods brought about by the political system of the Qing Dynasty government and the increase of survival pressure caused by the loss of pastureland, the production methods in Inner Mongolia Autonomous Region changed from nomadic grazing to limited nomadic grazing to four-seasons rotational grazing, and finally from continuous grazing to semi-agricultural and semi-pastoralism or even pure agriculture in a period of more than two hundred years. However, the Mongolian steppe is vast, and this change in the northern part of Inner Mongolia and the center of the pastoral area is half a century late.

(3) Individual Household Responsibility System - Fundamental changes in grazing systems in pastoral areas. In 1983, when the Double Contracting System was introduced, the pastoralists were very supportive at that time, after all, each individual household could finally have their own livestock and pasture, and became the owners in the real sense. In areas with large populations and small pastures, one of the inevitable results induced by the implementation of Grassland Contracting System Policy is that there occurred an inequitable use of pastures between pastoralists with more livestock and those with fewer livestock. After 1986, Ar Horqin Banner, Bairin Right Banner, Bordered Yellow Banner, Otog Banner, Urad Middle Banner and other areas with more people and less pastureland successively leased individual household pastureland. From Sep-

Chapter 2　Study Area

tember 1 to 5, 1989, the People's Government of Autonomous Region held a site meeting in Ar Horqin Banner on the system of leasing individual household pastureland, site visits to study the practices and experience of Ar Horqin Banner, exchange information about the situation with various places, as well as to establish and promote grass pasture "Individual Household Responsibility System". This system, together with the policy that allowed individual households to own their livestock and determine the price of their own livestock but communally shared the pastureland, are collectively known as the "Double Contracting System". In January 1990, the Party Committee and People's Government of the Autonomous Region pointed out that their intention to establish the pastureland contracting policy in the next three to five years in "Opinions on further deepening the reform of rural pastoral areas." According to statistics, in 1994, about 7,200 hectares of grassland in 25 provinces, cities and autonomous regions implemented different forms of contracting systems, accounting for 32% of the available grassland in China.

From September 6 to 10, 1994, the Department of Animal Husbandry and Veterinary Medicine of the State Ministry of Agriculture held an on–site national meeting on the implementation of paid contract use of pastureland in Ar Horqin Banner, Inner Mongolia Autonomous Region, which was attended by 28 provinces, cities and autonomous regions, including Inner Mongolia Autonomous Region, Qinghai Province and Xinjiang Uygur Autonomous Region. The meeting spoke highly of the experience of Ar Horqin Banner and requested that all parts of the country take this meeting as a new starting point for promoting the implementation of the pastureland contract system, increasing efforts, and striving to make greater progress in a shorter period of time. The meeting also pointed out that Ar Horqin Banner had created, summarized, and accumulated a more complete and more mature practical experience in the implementation of the pastureland contract system, hence worthy to be studied and learned from.

On October 27, 1994, the French newspaper Eurotimes published a report on "China fully implements pastureland contracting system, changing the pastureland management of livestock husbandry". The report highlighted the on‑site meeting held by the Chinese Ministry of Agriculture, which decided to promote the experience of Ar Horqin Banner of Inner Mongolia Autonomous Region on the pastureland contracting system by applying on the implementation of the same pastureland contracting system to the 400 million hectares of grass pastures in the country within three or four years. At this point, the Double Contracting System was established and fully promoted in the pastoral areas of the country, and at the same time, the nomadic system that had been inherited for thousands of years ended up in China, and the grassland nomadic grazing system had turned to sedentary and continuous grazing system.

The Double Contracting System had indeed produced obvious effects in motivating pastoralists to graze their livestock, mainly in the following two aspects. First, it was the rapid growth in the number of livestock. Taking Xilingol League, the main pasturing area in Inner Mongolia Autonomous Region as an example, the whole league had been striving from 1949 to 1966 to reach 10 million heads of livestock, but the goal has never been able to be achieved. And when the Individual Household Responsibility System was implemented in 1983, it took the league six years only that the total number of livestock exceeded the 10 million mark, fulfilling the dream of several generations. Second, the pastoralists' enthusiasm for capital construction reached unprecedented levels. From the 1980s onwards, under the guidance of government projects and policies, net fences began to be introduced in the grassland, and later, the pastoralists started to build houses, sheds, wells, fodder production sites, etc. By the end of the 1990s, more than 90% of pastoring households in the league completed basic construction, and the pastoralists basically settled in the pastoral areas of Inner Mongolia Autonomous Region.

2.2.3 Comparison of Grazing Systems in China and Mongolia

The policy of Individual Household Responsibility System in Inner Mongolia Autonomous Region is based on the principle of leasing pasture according to local conditions. The division of the pastureland took the village as a unit, and proposed a reasonable average area for each household according to the population, the number of livestock, pasture grade, grazing habits and other factors. The proposal would be discussed by the villagers committee and after they approved, they would lease the assigned pastureland to individual households for a period of usually 50 years. Meanwhile, in Mongolia, the pastoralists decide whether to conduct a four - seasons, three - seasons, or two - seasons rotation grazing based on the pasture conditions, and seasonal changes.

Mongolia's four - season rotational grazing system is the management system of Mongolian pastoral areas in which grassland is communally owned and livestock are privately owned. The four - season rotational grazing method is to rotate grazing areas of livestock within the pastoralist's pasture according to the seasonal climatic conditions, pasture conditions, water sources and livestock grazing habits. The difference between four - seasons rotational grazing and rotational grazing are that, firstly, there is no fence to isolate the pasture; secondly, the grazing time and resting time of four - seasons rotational grazing are mainly determined flexibly according to geographical conditions, seasonal changes, precipitation, pasture and livestock conditions, unlike the grazing time and resting time of rotational grazing which are mechanically fixed; thirdly, there is a certain distance between the use of pastureland in different seasons (such as summer and winter), which is generally in the range of 10 - 30km. A comparison of the evolution of grazing systems in China and Mongolia is shown in the Table 2 - 4 below.

Table 2-4 A comparison of the evolution of grazing systems in China and Mongolia

System	Mongolia		Inner Mongolia	
	1958 – 1990	1990 – present	1978 – 1990	1990 – present
Livestock ownership	Public	Private	Private	Private
Pasture ownership	Public	Public	Public	Private
Grazing system	Four – seasons rotational grazing	Four – seasons rotational grazing	Four – seasons rotational grazing	Continuous grazing

2.3 The Overview of the Study Site

The study site is located in a temperate typical grassland in the cross – border area of China and Mongolia, administratively under the Naran Soum of Sukhbaatar Province, Mongolia, and Naren Soum of Abag Banner, Xilingol League, Inner Mongolia Autonomous Region, China, both of which are purely pastoral areas. The vegetation types, climate, topography, soils, production methods (grazing) and stock rate are basically the same between the two. The soil is made up of chestnut soil with a humus layer of 5~10 cm thick, average annual precipitation from 1971 to 2016 was 220.6 ± 65.2mm whereby 60% to 80% of rain fell during the growing season (July to September) and the evaporation rate was 1505 ± 45.4mm (Table 2 – 5).

In study area, the foundation species are feathergrass (Stipa grandis) and Chinese rye grass (Leymus chinensis) while the dominant species are needle leaf sedge (Carex duriuscula), needle grass (Stipa sareptana), prairie sagewort (Artemisia frigida), Chenopodium acuminatum, Cleistogenes squarrosa, and Allium polyrhizum. Field survey data show that the cross – border area of the typical steppe between China and Mongolia has a total of 44 plant species in 16 families. From the comparison of different grazing systems, it is found that the species composition in four – seasons rotational grazing area (38 species) is the richest, followed by continuous grazing area (35 species) and forbidden grazing area (29 species) (Table 2 – 6).

Table 2 – 5 Ecological factors in the study area

Factors	Naran Soum	Naren Soum
Annual mean temperature (℃)	1.41a	1.03a
Annual mean precipitation (mm)	216.80 (a)	224.49 (a)
Altitude (m)	1356.21a	1346.39b
Average annual evaporation (mm/y)	1505.14a	1498.52a
Average stocking rate (sheep unit/km^2)	42a	50a
Soil type	Chestnut soil	Chestnut soil
Soil humidity (VWC)	7.5%a	6.9%b

Note: The lowercase letters indicate the significant difference at 5% level.

Source: Meteorological Bureau of Xilingol League and National Agency for Meteorology and Environmental Monitoring of Mongolia.

Stocking rate = The total number of livestock (sheep unit) owned by the pastoralist where the quadrat is located/The total area of the grassland owned by the pastoralist; Livestock data and grassland area were mainly obtained from the annual statistics data, and then verified via on – site survey and local community interviews, of which all types of livestock are being converted to sheep as the unit according to China's sheep unit conversion standard, as follow: 1 camel = 7 sheep, 1 horse = 6 sheep, 1 cow = 5 sheep and 1 goat = 1 sheep; Soil moisture: 20cm soil moisture tested with portable TDR 6500 tester.

Table 2-6 Species composition in the study area

Species	RG	FG	CG	Species	RG	FG	CG
Carex duriuscula	1	1	1	*Convolvulus ammannii*	1	0	1
Leymus chinensis	1	1	1	*Potentilla bifurca*	1	1	1
Stipa sareptana	1	1	1	*Lepidium apetalum*	1	0	1
Chenopodium acuminatum	1	1	1	*Ptilotricum canescens*	1	0	1
Cleistogenes squarrosa	1	1	1	*Saussurea mongolica*	1	1	1
Artemisia frigida	1	1	1	*Iris tenuifolia*	1	0	1
Allium polyrhizum	1	1	1	*Dontostemon micranthus*	1	0	1
Salsola collina	1	1	1	*Cymbaria dahurica*	1	1	1
Stipa grandis	1	1	1	*Artemisia sieversiana*	1	0	1
Allium tenuissimum	1	1	1	*Achnatherum splendens*	0	1	1
Caragana stenophylla	1	1	1	*Asparagus gobicus*	1	1	1
Neopallasia pectinata	1	1	1	*Allium mongolicum*	1	0	1
Artemisia annua	1	1	0	*Artemisia dracunculus*	1	0	1
Caragana microphylla	1	0	1	*Veronica didyma*	1	0	0
Oxytropis microphylla	1	1	1	*Poa annua*	1	1	0
Atriplex sibirica	1	0	0	*Koeleria litvinowii*	0	1	0
Kochia prostrata	1	0	1	*Haplophyllum dauricum*	1	0	1
Torilis scabra	1	1	1	*Limonium bicolor*	0	1	0
Heteropappus altaicus	1	1	1	*Bupleurum sibiricum*	0	1	0
Agropyron cristatum	1	1	1	*Galium verum*	1	0	0
Ephedra sinica	1	1	1	*Chenopodium aristatum*	0	0	1
Allium ramosum	1	1	1	*Allium condensatum*	0	1	0

Note: "1" represents the corresponding species, "0" represents no corresponding species.

2.3.1 Sukhbaatar Province

Sukhbaatar Province is located in the southeast of Mongolia. With an area of 82,900 square kilometers, its southern part is bordering with the Xilingol League of Inner Mongolia Autonomous Region, China, with a border of more than 470 kilometers

Chapter 2 Study Area

long. On its west is the Dornogovi Province, its north the Khentii Province and Dornod Province and its east the Dornod Province.

There is no railroads in the Sukhbaatar Province but a Baxin Railroad is planned to be built (from Choibalsan City of Dornod Province to Baruun − Urt of Sukhbaatar Province to the boundary of Mongolia, Biqigetu to the boundary of China, Zhu'engadabuqi Port, to the West Ujimqin Banner, East Ujimqin Banner and Bairin Right Banner of Inner Mongolia Autonomous Region, and to Fuxin of Liaoning before arriving at the Jinzhou Harbor). Roads are the main form of Sukhbaatar Province's transportation, and all mineral resources are transported by large trucks (with a maximum capacity of 60 tons) to and from the ports or railroad city hubs. Animal husbandry is the main economic industry in the province. Sukhbaatar Soum and Asgat Soum breeds find and semi − fine wool sheep, Bayandelger Soum breeds cattle for both meat and milk and Erdenetsagaan Soum breeds Ujimqin sheep. The province has 14,700 square kilometers of arable land and grows mainly cereals, potatoes, and vegetables, with a harvest of about 20,000 tons annually.

The main enterprises of Sukhbaatar Province are gold mines, printing plants, food processing plants, construction material plants, leather processing plants, fuel power plants and flour mills. Freight transport mainly relies on the highway, all soums in the province have telephone lines, while some soums also have computer networks and internet connections.

2.3.2 The Abag Banner of Xilingol League

The Abag Banner is located in the central − north part of Xilingol League, Inner Mongolia Autonomous Region. It is located at longitude 113°27′ to 116°11′ and latitude 43°04′ to 45°26′. Xilinhot and East Ujimqin Banner are the neighbours on its east,

Plain Blue Banner on its south, Sonid Left Banner on its west and Mongolia country on its north, with a span of 175km long national border. The Abag Banner has a total area of 27,500 square kilometres with its length from north to south of about 260km and its width from west to east of about 110km. The word Abag, is a Mongolian language, which means "uncle". Because of the descendent of Belgutei, the Genghis Khan's brother, named the tribe Belgutei led as the Abag tribe, and the name is used up to now.

The terrain of Abag Banner is mainly composed of low mountainous hilly areas on the Mongolian plateau, with an average elevation of 1,127 meters and a maximum elevation of 1,648 meters, and the terrain slopes from northeast to southwest. Morphologically, it can be divided into four types: Low hills, lava terraces, high plains, and sandy areas, with no mountains or gullies, and no obvious undulations in the terrain. The total population of the banner in 2017 was 43,949, mainly dominated by Mongolian and Han Chinese ethnic groups, with 18,864 Han Chinese, accounting for 43% of the total population, and 23,381 Mongolians, accounting for 54% of the total population. Among them, the pastoral population was 21,233, accounting for 40% of the total population while the urban population was 21,805, accounting for 51% of the total population.

National Highway 207 and Provincial Highway 101 run through the banner, and the total mileage of the road network in the territory reaches 1,058km, which can reach Hohhot, Beijing, Ulanqab, Baotou, Erenhot, Ordos, Tongliao, Chifeng, Chengde and other cities, so the traffic conditions are very convenient.

All six soums and up to 54 gacha (village) or 76% of total gacha in Abag Banner have roads, which are connected to the main Provincial Highway 101. The Provincial Highway 101 has nine country roadsd and border roads as its branches, radiated out to each soums and gacha of the Abag Banner. The Changannaoer Coalfield Line has got the approval to start the construction of a railroad from the Xilinhot to Erenhot.

The gross regional product of Abag Banner in 2016 was about 4.9 billion yuan. Of which, the contribution of the primary sector to economic growth was 7.4%, and the total number of livestock for the year was about 1,465,567 heads. Among them, 1,233,758 were small-sized animals, and 231,809 were large-sized animals. The contribution of the secondary sector to economic growth was 82.6%. Of which, the value added by industry was 315.387 million yuan, an increase of 26.6% over year 2015. The contribution of the tertiary sector to economic growth was 10%. Of which, the total retail sales of consumer goods was 749 million yuan, an increase of 15.6% over year 2015. From the location of business units, the retail sales of social consumer goods in cities and towns amounted to 572.95 million yuan, accounting for 76.5% of the total retail sales of consumer goods, an increase of 18.9%; the retail sales of consumer goods in pastoral areas amounted to 176.05 million yuan, an increase of 6.3%.

Chapter 3 Research Design and Methodology

3.1 Research Methodology

The research in this book was mainly based on the relevant theoretical methods of geography, biology and remote sensing. On the premise of obtaining primary data, the research integrated theoretical and empirical research methods, focusing on the combination of field survey and satellite image data, natural factors analysis and human factors analysis, qualitative research and quantitative analysis, historical literature research and current situation investigation research, as well as natural science research methods and social and humanistic science research methods.

3.1.1 Field Survey Method

Fieldwork is the traditional dominant research method in geography, including mac-

Chapter 3 Research Design and Methodology

ro – and micro – survey. From year 2010 to 2016, the authors visited most places of the five provinces in Mongolia (Sukhbaatar Province, Selenge Province, Tov Province, Dornod Province and Omnogovi Province) and five leagues and cities in Inner Mongolia Autonomous Region, China (Xilingol League, Chifeng City, Hulunbuir City, Ulanqab City and Tongliao City), gained a comprehensive understanding of the basic overview of grassland pastoralism and the socio – economic development of pastoral areas in Mongolia and Inner Mongolia, China, and conducted in – depth and solid field surveys and quadrat surveys of typical gachas.

3.1.2 Remote Sensing Monitoring Method

The survey methods for grassland vegetation are generally divided into two types: Ground monitoring and remote sensing monitoring. Ground monitoring is mainly used to determine vegetation environmental information such as grassland vegetation growth, yield estimation and species change by measuring parameters such as height, cover, yield and species of grassland vegetation. Although ground surveys are time – consuming and labor – intensive, they can reflect the characteristics of vegetation communities in a more detailed and comprehensive manner. On the other hand, remote sensing monitoring has become another major method for environmental monitoring of grassland vegetation because it saves time, effort and can be conducted over large areas and multiple time periods (Bao et al., 2017). Vegetation community change is a complex physiological process and is influenced by a variety of factors. Some remotely sensed surface parameters closely related to vegetation community change can be used to characterize the overall condition of vegetation communities. This research combined the use of ground monitoring and remote sensing monitoring tools to bring out the advantages of each.

3.1.3 Questionnaire Survey Method

The questionnaire method is a basic research method in humanities disciplines of sociology, anthropology and economics, and is one of the important research methods in human geography. Through purposeful and targeted questionnaire design and survey, we can fully understand and grasp the real attitudes, desires, understanding and awareness of different social strata on some specific matters. The research in this book focused on a documentary survey of various types of members in the society, including pastoralists and cadres of different incomes, ages, ethnicities, and cultures in pastoral areas, so as to obtain as comprehensive understanding as possible of the perceptions of relevant members of Mongolian pastoral society about changes in the ecological environment of pastoral areas.

3.1.4 Historical Documentation Method

The historical documentary method the research used was to make full use of various historical documents to compare the ecological environment in different periods of history, then analyzed the characteristics of the evolution of the grazing system and ecological environment in Mongolia and Inner Mongolia Autonomous Region, and further explored the causes of the degradation of Mongolian grasslands.

3.1.5 Statistical Analysis Method

The plant community, meteorological, and livestock data were counted using Excel 2010 software; remote sensing data were processed and NDVI was calculated using ENVI 5.0 (© Harris, Boulder, USA) and ArcGIS 10.0 (© ESRI, RedLands, USA) software; NDVI, quadrat data, and meteorological and livesock data from the three re-

Chapter 3 Research Design and Methodology

gions in Inner Mongolia Autonomous Region, Mongolia, and forbidden grazing area were tested for random significance ($p < 0.05$) using R.

3.2 Experimental Setup

As the topography characteristics of Mongolian grassland are flat with larger ecological niches and single and continuous ecological landscape (Zhou, 2014), two soums (one on each side of the national border) adjacent to each other and have similar biotic and abiotic conditions, were chosen as the study area. Three belt transects (each 40km long, crossing Mongolia and Inner Mongolia Autonomous Region) perpendicular to the borderline, and seven belt transects (each 20km long, four in Mongolia and three in Inner Mongolia Autonomous Region) parallel to the borderline, were set up in the study area with the help of Google Earth and GPS in accordance with the principle of combining belt transects with quadrat sampling methods and balanced random distribution of quadrats. Among them, the belt transect closest to the border within Mongolia was set 5km from the border line, within the border's closed management area, which is a forbidden grazing area. The intersection points of the perpendicular and parallel belt transects became the marked points to set up three individual quadrats, at a distance of 150 m from one another. The field survey was carried out during the peak biomass period, which was from the end of July to Mid – August of 2016. A total of 61 quadrats of 1m x 1m each were surveyed. In each quadrat, the number of species, total individual density, total aboveground biomass, total coverage and average height, soil type and its humidity were measured and recorded (Figure 3 – 1).

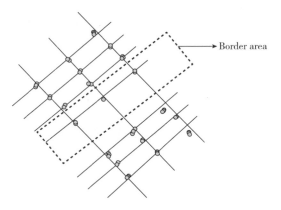

Figure 3 – 1　Spatial distribution of the quadrats

3.3　Quadrat Survey

3.3.1　General Characteristics of Plants

The plant community's characteristics were identified based on average height, total coverage, total individual density and total aboveground biomass (Zhang, 2004).

● Quadrat's average height: The natural mean height of the plant community in the quadrat was measured with measuring tape three times and the average value was taken.

● Quadrat's total coverage: The ratio of the vertical projection of the above-ground part of all available plant species to the quadrat's area. The average value of three persons' visual measurement results was taken.

Chapter 3 Research Design and Methodology

- Quadrat's total individual density: The total number of individuals for all species per quadrat, calculated via manual statistics method.
- Quadrat's total aboveground biomass: Calculated by harvesting all plants in the quadrats to measure their fresh weight with a 0.01g precision digital scale directly at the field.
- Quadrat's species richness: The number of species available in the quadrats.

3.3.2 Calculation of Species Dominance

The calculation of species dominance (IV) was as below:

$$IV(\%) = \frac{\text{Relative Coverage} + \text{Relative Density} + \text{Relative Frequency} + \text{Relative Height}}{4} \quad (3-1)$$

Where,

$$\text{Relative Density} = \frac{\text{Then umber of individuals of a species}}{\text{The number of individuals of all species}} \times 100 \quad (3-2)$$

$$\text{Relative Height} = \frac{\text{The average height of a species}}{\text{The sum of the average height of all species}} \times 100 \quad (3-3)$$

$$\text{Relative Coverage} = \frac{\text{The coverage of a species}}{\text{The sum of the coverage of all species}} \times 100 \quad (3-4)$$

$$\text{Relative Freqeuncy} = \frac{\text{The frequency of a species}}{\text{The frequency of all species}} \times 100 \quad (3-5)$$

3.3.3 Calculation of Species Diversity

Species diversity is the center of biodiversity and the most important structural and functional unit of biodiversity. It refers to the species abundant in animals, plants, microorganism etc. on earth. Species diversity includes two aspects: ①refers to the species richness in a certain area and can be called the regional species diversity; ②in ecolo-

蒙古草原放牧制度效益评价研究
Evaluation Research on the Benefits of Grazing System in the Mongolian Grassland

gy, it refers to the degree of evenness in species distribution and can also be referred to as biodiversity or community diversity. Species diversity is an objective index in measuring the abundance of biological resources in a certain area (Hurlbert, 1971).

When measuring the species diversity of the regional habitats, the absolute number of species in the community is usually compared so that the result is more concise. Even though calculating the number of all species within a quadrat to express the species richness reflects the number of species in the community, it omits the difference in the size of the contribution of dominant and rare species to diversity and is susceptible to the influence of field sampling area. Therefore, the result can be more accurate only by also applying Simpson dominance index, Shannon – Wiener index and Pielou's evenness index (Robert T. et al., 1995). The Shannon – Wiener diversity index is a parameter of the degree of species diversity and heterogeneity within the community, which can synthesize the species richness and evenness of communities, but ignores the species composition factor. Simpson dominance index is contrary to diversity and evenness index as the index reflects the changes in the number of species. The larger the value, the more unevenly distributed the number of species within the community and the more prominent the dominant species are. Pielou's evenness index reflects the degree of evenness in distribution of the number of individual species within various communities (Dickman & Mike, 1968).

Based on the number of plant species enumerated in the quadrat, the individual number and dominance index for all plant species, the diversity index of the community were calculated as below:

Species Richness $R = S$ (3 – 6)

Shannon – Wiener Index: $H' = -\sum_{n=1}^{\infty} P_i \ln(P_i)$ (3 – 7)

Chapter 3 Research Design and Methodology

Simpson's Diversity Index: $D = 1 - \sum_{n=1}^{\infty} (P_i)^2$ (3-8)

Pielou's Evenness Index: $\text{Pielou} = \dfrac{-\sum_{n=1}^{\infty} P_i \ln(P_i)}{\ln(S)}$ (3-9)

Where, R is the species richness, S is the number of species, P_i is the proportion of dominance of species i to the total dominance in the community.

3.3.4 Classification of Plant Groups

Plant functional group refers to a group of species or taxa that have similar responses under specific environmental factors. It is classified based on their biological, morphological, life history or other biological characteristics that are relevant to an ecosystem processes and to the behavior of the species (Griffin, 1988). Functional group is a basic unit to study the changes of plants according to its environment (Pérezharguindeguy et al., 2013). It is also an important unit for the study of biodiversity and the role of ecosystem function (Lavorel et al., 1997).

Therefore, this study selected a typical steppe located at the China – Mongolia border area as a study area to analyze the dominance of the ecological functional group and life – form functional group. Ecological functional group includes xerophytes, intermediate xerophytes, mesophytes and intermediate mesophytes while life – form functional group includes perennial grass, perennial weeds, annual grass as well as shrubs and sub – shrubs.

3.3.5 Normalized Difference Vegetation Index (NDVI)

It has been shown that the vegetation index can be used to show the vegetation growth condition, and the larger the value, the more photosynthetically active radiation (PAR) absorbed by the vegetation, the better the vegetation growth and the better vegetation community as a whole.

The vegetation has its own spectrum features, ie. strong absorption of visible lights and strong reflection of near – infrared (NIR) light. As there is a significant correlation between PAR spectral region and NIR spectral region (Na et al., 2010), red band and NIR band can be used to calculate vegetation index to reflect the vegetation growth condition. Examples of common vegetation indexes are Normalized Difference Vegetation Index, NDVI (Fan et al., 2009; Qin, 2019), Vertical Vegetation Index (Li et al., 2007), Ratio Vegetation Index, RVI (11), Difference Vegetation Index, DVI (Xue et al., 2004), Soil – adjusted Vegetation Index, SAVI, and Enhanced Vegetation Index, EVI (Huete, 1988). The vegetation indices commonly used to study the vegetation growth in typical grassland areas of arid – semi – arid are NDVI and Modified Soil – adjusted Vegetation Index, MSAVI (Wu et al., 2015).

The calculation method of NDVI is shown below:

$$NDVI = \frac{P_{NIR} - P_{Red}}{P_{NIR} + P_{Red}} \quad (3-10)$$

Where, NDVI is Normalized Difference Vegetation Index, P_{NIR} is a near – infrared band and P_{Red} is a red band.

NDVI is a type of vegetation index which has the highest correlation with the greenness indices of herbaceous plants (Carlson & Ripley, 1997). When vegetation coverage is 25% – 80%, the NDVI value increases linearly with vegetation coverage; when the vegetation coverage is larger than 80%, the monitoring sensitivity decreases (Meng, 2006). Meanwhile, NDVI is more sensitive to the changes in soil and is suitable for arid area vegetation survey and monitoring during their early and middle growth phases. The vegetation cover of the study area in this book is basically less than 80%, and the comparison between NDVI results and MSAVI results reveals that NDVI better

Chapter 3 Research Design and Methodology

reflects the difference in vegetation cover between the three comparison areas. Therefore, the study in this book used the Normalized Difference Vegetation Index (NDVI).

Five periods of cloudless satellite images from the study area (126/29), obtained from Thematic Mapper (TM) and Enhanced Thematic Mapper (ETM), were utilized for the study (http://earthexplorer.usgs.gov/). The image resolution was 30 m and Band 3 (0.66μm) and Band 4 (0.84μm) were mainly used. From these images, the dynamic changes of vegetation were identified through a series of image processing, i.e. geometric correction, atmospheric correction, radiometric calibration, NDVI calculation, clipping and statistical calculation (Table 3–1).

Table 3–1 Remote sensing data source details

Time	Path/Row	Band Information	Satellite/Sensor	Resolution
1989/8/3	126/29	B3 (0.66) /B4 (0.84)	Landsat5/TM	30m
1993/9/15	126/29	B3 (0.66) /B4 (0.84)	Landsat5/TM	30m
2005/7/14	126/29	B3 (0.66) /B4 (0.84)	Landsat5/TM	30m
2011/7/31	126/29	B3 (0.66) /B4 (0.84)	Landsat5/TM	30m
2016/8/13	126/29	B3 (0.66) /B4 (0.84)	Landsat8/ETM+	30m

In order to obtain an accurate NDVI value for the above 61 quadrats, a total of 9 pixels, ie. the pixel located at the center of every quadrat and the eight surrounding pixels were included in the calculation to obtain the average NDVI value.

3.3.6 Calculation of Plant Community Stability

Since Macarthur proposed the diversity – stabilityhypothesis in the 1950s, the issues of diversity and stability have always been a debating topic (Macarthur, 1955). Therefore, Boucot (1985) analyzed and pointed out that the reason for two conflicting hypotheses of diversity – stability relationship are due to diverse definitions of diversity, complexity and stability of ecology. Regarding the community diversity and stability, scholars have carried out many research works, which are supported by a considerable number of phenomena and logical reasoning, and have inspired many (Mcnaughton, 1991; Tilman & Haddi, 1992).

Grassland ecosystems have dissipative structural features (Zhou, 1989). Based on the dissipative structural hypothesis, the ecosystem achieves an orderly harmony through the interaction between function ↔ structure ↔ fluctuations. The "fluctuation" mentioned here can be known as an ecological phenomenon, which refers to the deviation of the system from stability under the influence of internal factors or external factors. Fluctuation is the lever that triggers the change of ecological order and the change will inevitably lead to the change of stability. Therefore, the fluctuation is closely related to the structure and stability of the system. Stability is referred to the resiliency of the ecosystem returning to its original state after disturbance and is usually measured mathematically or empirically.

M. Godron stability test is a method discovered by French ecologist from industrial production and introduced into plant ecology. It is a method to calculate the stability from the number and frequency of all species in a plant community (Godron et al., 1971). The study of this book used the M. Godron stability test to calculate plant community stability in the study area and to explore the stability of vegetation communities under different grazing systems with the purpose of providing new ideas to reveal the stability mechanism of grassland. The method of M. Godron stability test was as follow:

i. The frequency measurements of different plant species in the studied community were arranged in descending order and the plant species were placed in correspondence with their frequencies.

Chapter 3   Research Design and Methodology

ii. The plant's frequency was converted into relative frequency and the cumulative value in descending order was calculated.

iii. The ratio of the corresponding plant species to the total species was calculated.

iv. The intersection point between the equation of the smooth curve obtained from the simulated scatter and y = 1 − x was used to determine the cumulative relative frequency and the proportion of the corresponding plant species to the total number of species coordinates x/y. 0.2/0.8 was the stability point of the community, the smaller the distance (d_x) between the intersection point coordinates x/y and 0.2/0.8, the more stable the community was.

The natural conditions of the study area are basically the same, the space span of the study area was not too large and the difference in plant species' frequency was not too big. Therefore, vegetation cover stability was further calculated based on the calculation of frequency stability, while comparing the sensitivity of frequency stability in the study area and highlighting the differences of communities in different grazing areas.

We revised M. Godron's stability test to determine the vegetation stability in different grazing areas by replacing the frequency of various plants with coverage in the test. The revised method of M. Godron stability test was as follow:

i. The coverage measurements of different plant species in the studied community were arranged in descending order and the plant species were placed in correspondence with their coverage.

ii. The plant's coverage was converted into relative coverage and the cumulative value in descending order was calculated.

iii. The ratio of the corresponding plant species to the total species was calculated.

iv. The intersection point between the equation of the smooth curve obtained from the simulated scatter and y = 1 − x was used to determine the cumulative relative frequency and the proportion of the corresponding plant species to the total number of species coordinates x/y. 0.2/0.8 was the stability point of the community, the smaller the distance (d_x) between the intersection point coordinates x/y and 0.2/0.8, the more stable the community was.

Chapter 4   Analysis of Results

4.1   General Characteristics of Plant Community

The general characteristics of the plant community included the community average height, total coverage, total individual density and total aboveground biomass.

Under different grazing systems, community average height reduced from forbidden grazing (FG), 21.2 cm > rotational grazing (RG), 14.8 cm > continuous grazing (CG), 8.4 cm, with a significant difference among the three systems ($p < 0.05$) (Figure 4 – 1).

Total coverage reduced from forbidden grazing (67.9%) > rotational grazing (64.3%) > continuous grazing (56.5%), with a significant difference between rotational grazing and continuous grazing ($p < 0.05$) but not between rotational grazing and forbidden grazing as well as forbidden grazing and continuous grazing (Figure 4 – 2).

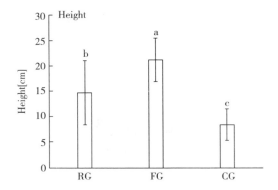

Figure 4 – 1　Community average height in different grazing system areas

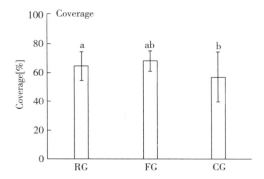

Figure 4 – 2　Community total coverage height in different grazing system areas

Total aboveground biomass reduced from forbidden grazing, 455.9 g/m^2 > rotational grazing, 268.4 g/m^2 > continuous grazing, 122.2 g/m^2, with a significant difference among the three systems ($p < 0.05$) (Figure 4 – 3).

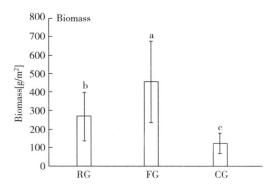

Figure 4 – 3　Community total aboveground biomass in different grazing system areas

Chapter 4 Analysis of Results

Total individual density decreased from rotational grazing, 439.4 individuals/m^2 > continuous grazing, 310.6 individuals/m^2 > forbidden grazing 228.4 individuals/m^2, among which, there was a significant difference between rotational grazing and both continuous grazing and forbidden grazing ($p < 0.05$) but not between continuous grazing and forbidden grazing (Figure 4 – 4).

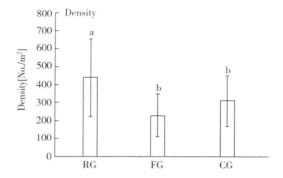

Figure 4 – 4 Community total individual density in different grazing system areas

4.2 Analysis of Species Dominance

Species dominance can indicate the relative importance of each plant in the community and the optimum habitat of the plant. The changes in dominance can affect the community structure, ie. the higher the species dominance, the more obvious the species is at advantage position (Zhang, 2004).

This book focused on the plant species with species dominance larger than 3% in the entire study area, and found that needleleaf sedge (*Carex duriuscula*), Chinese rye grass (*Leymus chinensis*), *Stipa sareptana*, *Chenopodium acuminatum*, *Cleistogenes squarrosa*, prairie sagewort (*Artemisia* frigida), *Allium polyrhizum*, slender Russian – thistle (*Salsola collina*), needlegrass (*Stipa grandis*) and *Allium tenuissimum* had high average dominance in the grassland community of different grazing system, in which the cumulative dominance of these species accounted for 72.46 % (Table 4 – 1).

Table 4 – 1 Plant species with an average dominance larger than 3% in the study area

Species	Dominance (%)	Order	Facilitated Grazing System
Carex duriuscula	16.11 ± 10.56	1	RG
Leymus chinensis	10.99 ± 9.94	2	FG
Stipa sareptana	8.11 ± 9.40	3	CG
Chenopodium acuminatum	7.36 ± 6.41	4	CG
Cleistogenes squarrosa	6.62 ± 5.63	5	CG
Artemisia frigida	6.32 ± 5.30	6	RG
Allium polyrhizum	5.47 ± 6.26	7	CG
Salsola collina	4.66 ± 4.03	8	CG
Stipa grandis	3.66 ± 8.34	9	FG
Allium tenuissimum	3.16 ± 2.90	10	CG
Total	72.46	—	—

The 10 species with a dominance greater than 3% in the study area differed in their effects on different grazing systems, could be categorized into three types (Table 4 – 2): rotational grazing (two species), forbidden grazing (two species) and continuous grazing (six species).

Chapter 4 Analysis of Results

Table 4-2 Dominance of major species for different grazing systems %

Species	RG	FG	CG
Carex duriuscula	22.27 ± 10.24a	10.67 ± 10.66b	11.37 ± 7.45b
Leymus chinensis	11.15 ± 13.23a	15.91 ± 6.44a	9.56 ± 5.91a
Stipa sareptana	3.46 ± 5.08b	2.84 ± 5.07b	14.13 ± 10.23a
Chenopodium acuminatum	6.34 ± 5.65a	6.58 ± 7.77a	8.57 ± 6.8a
Cleistogenes squarrosa	5.52 ± 4.18b	3.57 ± 2.5b	8.51 ± 6.85a
Artemisia frigida	6.68 ± 6.34a	5.61 ± 5.37a	6.15 ± 4.2a
Allium polyrhizum	5.34 ± 7.99a	5.01 ± 3.84a	5.71 ± 4.81a
Salsola collina	4.15 ± 3.71a	2.44 ± 3.1a	5.75 ± 4.32a
Stipa grandis	3.27 ± 6.93b	12.8 ± 13.68a	1.68 ± 6.53b
Allium tenuissimum	1.99 ± 2.33b	3.59 ± 2.5ab	4.23 ± 3.15a

(1) The dominance species facilitated by rotational grazing system:

The dominance of *C. duriuscula* and *A. frigida* in different grazing areas reduced from rotational grazing area > continuous grazing area > forbidden grazing area.

The dominance of *C. duriuscula* in rotational grazing area, forbidden grazing area and continuous grazing area were 22.27%, 10.67% and 11.37% respectively. The significant difference of its dominance was greater in the rotational grazing area than in both forbidden grazing area and continuous grazing area ($p < 0.05$) but there was no significant difference between forbidden grazing area and continuous grazing area.

The dominance of *A. frigida* in rotational grazing area, forbidden grazing and continuous grazing area were 6.68%, 5.61% and 6.15% respectively. There was no significant difference among the three types of grazing areas.

(2) The dominance species facilitated by forbidden grazing system:

The dominance of *L. chinensis* and *S. grandis* in different grazing areas reduced from forbidden grazing area > rotational grazing area > continuous grazing area.

The dominance of *L. chinensis* in rotational grazing area, forbidden grazing area and continuous grazing area were 11.15%, 15.91% and 9.56% respectively. There was no significant difference among the three different types of grazing areas.

The dominance of *S. grandis* in rotational grazing area, forbidden grazing area and continuous grazing area were 3.27%, 12.8% and 9.56% respectively. The significant difference of its dominance was greater in the forbidden grazing area than in both rotational grazing area and continuous grazing area ($p < 0.05$) but there was no significant difference between rotational grazing area and continuous grazing area.

(3) The dominance species facilitated by continuous grazing system:

The dominance of *S. sareptana*, *C. acuminatum*, *C. squarrosa*, *A. polyrhizum*, *S. collina* and *A. tenuissimum* in continuous grazing area was larger than in both rotational grazing area and forbidden grazing area. Among which, the dominance of *S. sareptana*, *C. squarrosa*, *A. polyrhizum* and *S. collina* in different grazing areas reduced from continuous grazing area > rotational grazing area > forbidden grazing area (none of these species were found in forbidden grazing area); while the dominance of *C. acuminatum* and *A. tenuissimum* in different grazing areas reduced from continuous grazing area > forbidden grazing area > rotational grazing area.

The dominance of *S. sareptana* in rotational grazing area, forbidden grazing area, and continuous grazing area are 3.46%, 2.84% and 14.13% respectively. The significant difference of its dominance was greater in the continuous grazing area than in both rotational grazing area and forbidden grazing area ($p < 0.05$) but there was no significant difference between rotational grazing area and forbidden grazing area.

The dominance of *C. squarrosa* in rotational grazing area, forbidden grazing area and continuous grazing were 5.52%, 3.57% and 8.51% respectively. The significant difference of its dominance was greater in the continuous grazing area than in both rotational grazing area and forbidden grazing area ($p < 0.05$) but there was no significant difference between rotational grazing area and forbidden grazing area.

The dominance of *A. polyrhizum* in rotational grazing area, forbidden grazing area, and continuous grazing area were 5.34%, 5.01% and 5.71% respectively. There was no significant difference among the three different types of grazing areas.

The dominance of *S. collina* in rotational grazing area, forbidden grazing area, and continuous grazing area were 4.15%, 2.44% and 5.75% respectively. There was no significant difference among the three different types of grazing areas.

The dominance of *C. acuminatum* in rotational grazing area, forbidden grazing area, and continuous grazing area were 6.34%, 6.58% and 8.57% respectively. There was no significant difference among the three different types of grazing areas.

The dominance of *A. tenuissimum* in rotational grazing area, forbidden grazing area, and continuous grazing area were 1.99%, 3.59% and 4.23% respectively. The significant difference of its dominance was greater in the continuous grazing area than in rotational grazing area ($p < 0.05$) but there was no significant difference between continuous grazing area and forbidden grazing area, as well as between rotational grazing area and forbidden grazing area.

4.3 Analysis of Species Diversity

Species richness, R, refers to the number of plant species in the community and is one of the most direct and effective methods to depict the species diversity. Under different grazing systems, the R of rotational grazing area, forbidden grazing area and continuous grazing area were 12.37 species/m^2, 11.43 species/m^2 and 12.52 species/m^2

respectively. The R value reduced from continuous grazing area > rotational grazing area > forbidden grazing area, with no significant difference among the three different types of grazing areas (Figure 4 – 5).

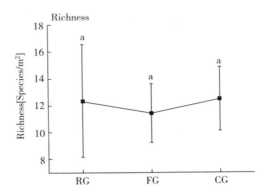

Figure 4 – 5　Species richness for three different types of grazing areas

Shannon – Wiener index, also known as the information index, reflects the amount of information on the diversity of plant communities. Under different grazing systems, the Shannon – Wiener index for rotational grazing area, forbidden grazing area and continuous grazing area were 2.19, 2.22 and 2.28 respectively. The index value reduced from continuous grazing area > forbidden grazing area > rotational grazing area, with no significant difference among the three different types of grazing areas (Figure 4 – 6).

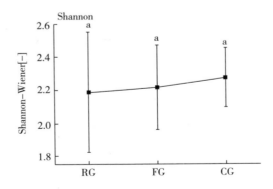

Figure 4 – 6　Shanner – Wienner index for three different types of grazing areas

Simpson diversity index reflects the degree of differentiation of the species quantity. Under different grazing systems, the Simpson index for rotational grazing area, forbidden grazing area and continuous grazing area were 0.85, 0.86 and 0.87 respectively. The index reduced from continuous grazing area > forbidden grazing area > rotational grazing area. The significant difference of the index was greater in continuous grazing area and forbidden grazing area than in rotational grazing area but there was no significant difference between continuous grazing area and forbidden grazing area ($p < 0.05$) (Figure 4 - 7).

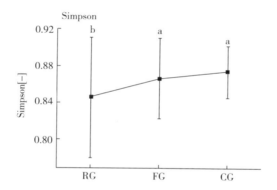

Figure 4 - 7 Simpson diversity index for three different types of grazing areas

Pielou's evenness index refers to the proportional distribution of the number of individuals of each species in the community. Under different grazing systems, the index for rotational grazing area, forbidden grazing area and continuous grazing area were 0.89, 0.92 and 0.91 respectively. The index reduced from forbidden grazing area > continuous grazing area > rotational grazing area, with no significant difference among the three different grazing areas (Figure 4 - 8).

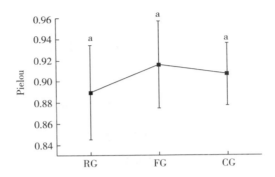

Figure 4 – 8 Pielou's evenness index for three different types of grazing areas

4.4 Characterization of Plant Functional Groups

Table 4 – 3 showed that under different grazing systems, the dominance of xerophytes in rotational grazing area, forbidden grazing area and continuous grazing area are 55.6, 66.76 and 68.21 respectively. Its dominance reduced from continuous grazing area > forbidden grazing area > rotational grazing area. The significant difference of its dominance was greater in both continuous grazing area and forbidden grazing area than in rotational grazing area but there was no significant difference between continuous grazing area and forbidden grazing area.

The dominance of intermediate xerophytes in rotational grazing area, forbidden grazing area and continuous grazing area were 27.39, 15.37 and 13.65 respectively. Its dominance decreased from rotational grazing > forbidden grazing > continuous grazing. The significant difference of its dominance was greater in rotational grazing area than in both forbidden grazing area and continuous grazing area but there was no significant difference between forbidden grazing area and continuous grazing area.

The dominance of intermediate mesophytes in rotational grazing area, forbidden grazing area and continuous grazing area are 8.78, 10.17 and 11.62 respectively. Its dominance decreased from continuous grazing area > forbidden grazing area > rotational grazing area and there was no significant difference among the three different types of grazing areas.

The dominance of mesophytes in rotational grazing area, forbidden grazing area and continuous grazing area were 8.23, 7.69 and 6.12 respectively. Its dominance decreased from rotational grazing area > forbidden grazing area > continuous grazing area and there was no significant difference among the three different types of grazing areas.

Table 4-3 The dominance of water ecological groups in different types of grazing areas

		Mean ± SD			Mean ± SD
Xerophytes	RG	55.6 ± 11.82b	Intermediate mesophytes	RG	8.78 ± 6.31a
	FG	66.76 ± 15.99a		FG	10.17 ± 10.13a
	CG	68.21 ± 13.46a		CG	11.62 ± 8.96a
Intermediate xerophytes	RG	27.39 ± 9.56a	Mesophytes	RG	8.23 ± 6.72a
	FG	15.37 ± 10.03b		FG	7.69 ± 4.04a
	CG	13.65 ± 7.3b		CG	6.12 ± 7.94a

Note: SD refers to standard deviation.

Table 4-4 showed that the dominance of perennial grass in rotational grazing area, forbidden grazing area and continuous grazing area were 24.62, 38.16 and 36.07 respectively. Its dominance decreased from forbidden grazing area > continuous grazing area > rotational grazing area. The significant difference of its dominance was greater in both forbidden grazing area and continuous grazing area than in rotational grazing area but there was no significant difference between continuous grazing area and forbidden grazing area.

Table 4-4 The dominance of life-form functional groups in different types of grazing areas

		Mean ± SD			Mean ± SD
Perennial grass	RG	24.62 ± 11.76b	Annual grass	RG	24.61 ± 11.68a
	FG	38.16 ± 17.27a		FG	15.32 ± 17.88a
	CG	36.07 ± 13.15a		CG	19.49 ± 10.79a
Perennial weeds	RG	43.28 ± 16.32a	Shrubs and sub-shrubs	RG	7.48 ± 5.29a
	FG	36.89 ± 8.96ab		FG	9.63 ± 3.94a
	CG	35.65 ± 11.71b		CG	8.4 ± 5.69a

Note: SD refers to standard deviation.

The dominance of perennial weeds in rotational grazing area, forbidden grazing area and continuous grazing area were 43.28, 36.89 and 35.65 respectively. Its dominance decreased from rotational grazing area > forbidden grazing area > continuous grazing area. The significant difference of its dominance was greater in rotational grazing area than in both forbidden grazing area and continuous grazing area but there was no significant difference between forbidden grazing area and continuous grazing area.

The dominance of annual grass in rotational grazing area, forbidden grazing area and continuous grazing area were 24.61, 15.32 and 19.49 respectively. Its dominance decreased from rotational grazing area > continuous grazing area > forbidden grazing area and there was no significant difference among the three different types of grazing areas.

The dominance of shrubs and sub-shrubs in rotational grazing area, forbidden grazing area and continuous grazing area were 7.48, 9.63 and 8.4 respectively. Its dominance decreased from forbidden grazing area > continuous grazing area > rotational grazing area and there was no significant difference among the three different types of grazing areas.

4.5 Normalized Difference Vegetation Index (NDVI)

Figure 4 – 9 showed that the NDVI values of 1989 in rotational grazing area, forbidden grazing area and continuous grazing area were 0.06, 0.11 and 0.04 respectively. The values decreased from forbidden grazing area > rotational grazing area > continuous grazing area. The significant difference of the value was greater in the forbidden grazing area than in both rotational grazing area and continuous grazing area ($p < 0.05$) but there was no significant difference between rotational grazing area and continuous grazing area.

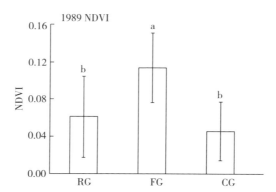

Figure 4 – 9 The NDVI values of 1989 in three different types of grazing areas

Figure 4 – 10 showed that the NDVI values of year 1993 in rotational grazing area, forbidden grazing area and continuous grazing area were 0.11, 0.16 and 0.12 respectively. The values decreased from forbidden grazing area > continuous grazing area > rotational grazing area. The significant difference of the value was lower in rotational grazing than in forbidden grazing area (p < 0.05) but there was no significant difference between continuous grazing area and both rotational grazing area and forbidden grazing area.

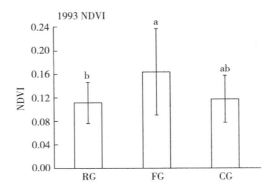

Figure 4 - 10　The NDVI values of 1993 in three different types of grazing areas

Figure 4 - 11 showed that the NDVI values of 2005 in rotational grazing area, forbidden grazing area and continuous grazing area are 0.04, 0.07 and 0.03 respectively. The values decreased from forbidden grazing area > rotational grazing area > continuous grazing area. The significant difference of the values was greater in the forbidden grazing area than in continuous grazing area (p < 0.05) but there was no significant difference between rotational grazing area and both forbidden grazing area and continuous grazing area.

Figure 4 - 12 showed that the NDVI values of 2011 in rotational grazing area, forbidden grazing area and continuous grazing area were 0.3, 0.29 and 0.17 respectively. The value decreased from rotational grazing area > forbidden grazing area > continuous grazing area. The significant difference of the value was greater in both rotational grazing

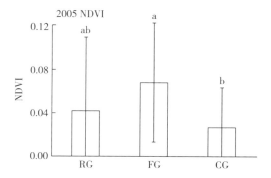

Figure 4 – 11 The NDVI values of 2005 in three different types of grazing areas

area and forbidden grazing area than in continuous grazing area ($p < 0.05$) but there was no significant difference between rotational grazing area and forbidden grazing area.

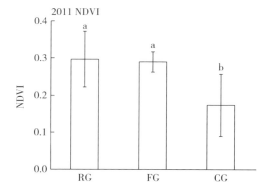

Figure 4 – 12 The NDVI values of 2011 for three different types of grazing areas

Figure 4 – 13 showed that the NDVI values of 2016 in rotational grazing area, forbidden grazing area and continuous grazing are 0.07, 0.16 and 0.04 respectively. The values decreased from forbidden grazing area > rotational grazing area > continuous grazing area and there was significant differences among the three different types of grazing areas ($p < 0.05$).

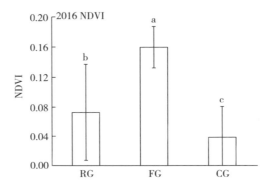

Figure 4-13 The NDVI values of 2016 in three different types of grazing areas

4.6 Plant Community Stability Index

Table 4-5 and Figure 4-14 showed the community stability result of three different grazing system areas based on frequency: the x/y and d_x of rotational grazing area were 0.36/0.66 and 0.21 respectively; the x/y and d_x of forbidden grazing area were 0.37/0.63 and 0.24 respectively; the x/y and d_x of continuous grazing area were 0.34/0.66 and 0.20 respectively. The frequency-based community stability index in continuous grazing area was the highest, followed by rotational grazing area and lastly, forbidden grazing area, ie. continuous grazing area > rotational grazing area > forbidden grazing area.

Table 4-5 The community stability based on frequency

GS	TOC	CC	PV	C (x/y)	d_x	SO
RG	$y = -1.2226x^2 + 2.138x + 0.0544$	0.9978	P<0.01	0.36/0.66	0.21	2
FG	$y = -1.109x^2 + 2.0344x + 0.0322$	0.9957	P<0.01	0.37/0.63	0.24	3
CG	$y = -1.3577x^2 + 2.2446x + 0.0568$	0.9854	P<0.01	0.34/0.66	0.20	1

Note: GS = Grazing system, TOC = Type of curves, CC = Correlation coefficient, PV = P value (the significant difference), C = Coordinate, SO = Stability order.

Chapter 4 Analysis of Results

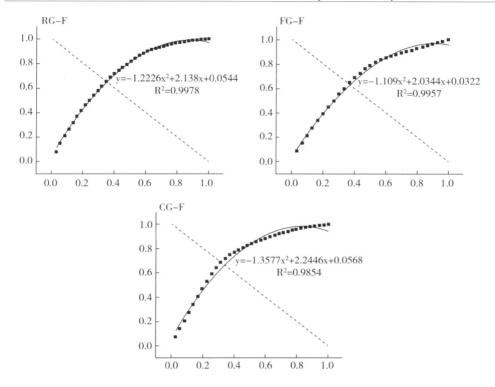

Figure 4 – 14 The graphs of frequency – based plant community stability in different grazing system areas

Table 4 – 6 and Figure 4 – 15 showed the community stability result of three different grazing system areas based on coverage: the x/y and d_x of rotational grazing were 0.26/0.74 and 0.09 respectively; x/y and d_x of forbidden grazing area were 0.31/0.70 and 0.16 respectively; the x/y and d_x of continuous grazing were 0.28/0.72 and 0.12 respectively. The coverage – based community stability index of rotational grazing area was the highest, followed by continuous grazing area and lastly, forbidden grazing area, ie. rotational grazing area > continuous grazing area > forbidden grazing area.

Table 4-6 The community stability based on coverage

GS	TOC	CC	PV	C (x/y)	d_x	SO
RG	$y = -1.1701x^2 + 1.7245x + 0.3847$	0.9294	P < 0.01	0.26/0.74	0.09	1
FG	$y = -1.5384x^2 + 2.3522x + 0.1201$	0.9854	P < 0.01	0.31/0.70	0.16	3
CG	$y = -1.2765x^2 + 1.9303x + 0.2887$	0.9729	P < 0.01	0.28/0.72	0.12	2

Note: GS = Grazing system, TOC = Type of curves, CC = Correlation coefficient, PV = P value (significant difference), C = Coordinate, SO = Stability Order.

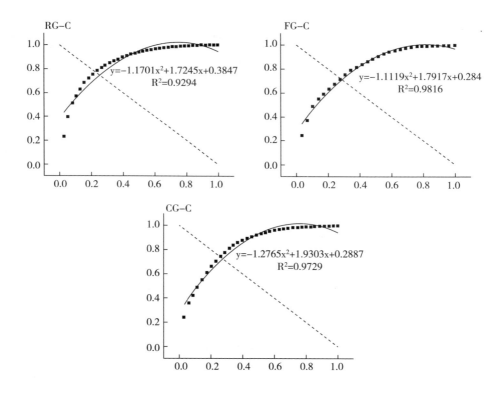

Figure 4-15 The graphs of coverage-based plant community stability in different grazing system areas

Chapter 5　Discussion

5.1　Effects of Different Grazing Systems on Plant Community Characteristics

As direct managers of the composition and diversity of grassland plant communities, the herbivores' foraging behavior is the most direct factor influencing changes in plant community population dynamics and diversity (Milchunas et al., 1988; Li, 1993; Han & Ritchie, 1998; Metea et al., 2010).

The average height, total coverage and total aboveground biomass of the plant community in forbidden grazing area were larger than in grazing areas (ie. rotational grazing area and continuous grazing area). The significant difference of the value of average height and total aboveground biomass was greater in forbidden grazing areas than in grazing areas (Figure 4 – 1, Figure 4 – 2 and Figure 4 – 3). Such results are the same as many other studies. The main reason may be that the foraging and trampling acts of grazing livestock reduce the leaf area index and plants' photosynthesis capacity, which changes the structure of pasture and thus affects the basic characteristics of plant communities (Li, 2002; Carrera et al., 2008; Zhao, 2010). Previous studies have shown that vegetation within forbidden grazing area which is not impacted by foraging and trampling of livestock, grow rampantly, hence, the average height, total aboveground biomass and species diversity are significantly greater than those in the grazing areas (Zhao, 2003; Firincioglu et al., 2007; Deak et al., 2016). However, some studies also pointed out that the dominant species in the forbidden grazing area will directly or indirectly control the vitality of other species. Therefore, intense competition between species will suppress or eliminate other weaker species, may lower the species diversity, reduce the number of shorter plants and eventually reduce the biomass (Grime, 1977; Collins et al., 1988; Loeser et al., 2007; Yan & Tang, 2007; Brinkert et al., 2015; Niu et al., 2018). The study in this book showed that there was no significant difference in total coverage between forbidden grazing area and both rotational grazing area and continuous grazing area; and the value of the total individual density decreased from rotational grazing area > continuous grazing area > forbidden grazing area; the value of species richness decreased from continuous grazing area > rotational grazing area > forbidden grazing area (Figure 4 – 2, Figure 4 – 4 and Figure 4 – 5). The above results explained that the average height, total coverage and total aboveground biomass of plant communities in grazing areas decrease but the removal of surface and old plant tissues can induce new plant growth by providing larger growing space for weaker species and thus, increase the total individual density and species diversity.

Even though the average height and total aboveground biomass of plant communities in forbidden grazing area were significantly greater than those in grazing areas, the cumulative growth of plant height and biomass of the plant communities in forbidden grazing area and grazing areas during the whole growing period needed to be set up for compensatory plant growth experiments to determine the superiority of each. Grazing can have both mechanisms to suppress and enhance plant growth; and the compensatory plant growth is subjected to the net effect between suppressing and enhancing events, where the net effect is closely related to the plant community type, grazing system, grazing intensity, environmental conditions, etc. (Mcnaughton & Oesterhelds, 1990).

There were significant differences in the average height, total coverage and total aboveground biomass of plant communities in rotational grazing area and continuous grazing area (Figure 4 - 1, Figure 4 - 2 and Figure 4 - 3). It is possibly because rotational grazing is based on the characteristic that forage grasses of different growing stages differ in their sensitivity to grazing (Bai et al., 2004; Yang, 2009). By maintaining grasslands above the minimum physiological growth threshold tolerated for sustainable function and health through human control, vegetation gradually returns to pre - grazing health levels during the sufficient regeneration period (Garcia et al., 2014). Ultimately, it can increase forage production and improve stocking rate (Severson & Martin, 1988). Grasslands damaged within their threshold can recover through their own physiological mechanisms, beyond which degradation can occur (Tilman & Haddi, 1992; Han, 1993; Peng & Wang, 2005; Hao et al., 2013).

5.2 Effects of Different Grazing Systems on the Dominance of Major Species

There are 44 species of plants in 16 families in the typical grassland of the study area under different grazing systems, among which Poaceae, Asteraceae, Liliaceae and Chenopodioideae accounted for 18.18%, 15.91%, 13.64% and 11.36% respectively, totaling about 60%; followed by Leguminosae and Brassicaceae, both accounted for 6.8% (Table 2-7).

Species dominance depicts vegetation change under grazing conditions and is more comprehensive and accurate than using single indexes such as plant or vegetation coverage and biomass (Wang & Liu, 1996). The dominance of *L. chinensis*, *S. sareptana*, *C. squarrosa*, and *S. grandis* was high grassland communities with different grazing systems. This is because long-term feeding and other behavioral disturbances induce changes in composition and structure of typical steppe's dominant species or dominant functional groups despite the differences in livestock feeding options, movement among different grazing systems. However, some perennial grasses such as *L. chinensis*, *S. sareptana*, *C. squarrosa*, *S. grandis* etc have strong adaptability. Although they are severely dwarfed in grazing areas, they have formed a typical grassland's unique perennial grass-dominated plant community (Kahmen & Poschlod, 2004; Fartmann et al., 2012). This proves that these dominant species can maintain their strong competitive advantage and are dominant over the grassland plant community.

Besides, the dominance of *C. duriuscula*, *A. frigida*, *A. polyrhizum*, and *A. tenuissimum* was relatively high in grazing areas with different grazing systems. This indicated that the degradative succession of grassland in the study area was obvious. The relatively high dominance of these species indicated that the degradation of grassland in the study area was serious, resulting in an increase of annual grasses in the grassland community.

Other than the succession due to grassland degradation, the selective intensity of grazing on grassland community species is also the key factor affecting species dominance. The impacts of grazing on plant communities will ultimately be reflected in the plant's composition and population structure. Grazing can affect the community structure from various aspects such as the number and growth of tiller buds (Olson & Richards, 1988), spatial distribution of tiller (Richards & Olsonj, 1998; Buttler & Briske; 1989), basal area of the plant (Anderson & Briskev, 1990), compensatory properties between tiller density and individual size (Matthew et al., 1995) etc.

Qualitative indicator plant refers to a plant that appears (positive) or disappears (negative) when the grazing intensity exceeds a certain threshold and its occurrence can indicate the intensity of grazing (Waldhardt & Otte, 2003).

Quantitative indicator plant refers to the dominance increased (positive) or decreased (negative) with grazing intensity, hence the degree of dominance can indicate the strength of grazing intensity. Positive indicator plant is the increaser while negative indicator plant is the decreaser (Dyksterhuis, 1949).

The results of the study showed that the dominance of *L. chinensis* and *S. grandis* was higher in the forbidden grazing area, of which the dominance of *S. grandis* in forbidden grazing area was significantly greater than in rotational grazing area and continuous grazing area. Such a result conforms to the studies of An (2001), Yang and his colleagues (2009) and Li (1993). Studies had pointed out that *L. chinensis* and *S. grandis* are negative qualitative indicator plants, in which their dominance showed a decreasing trend with the increase of grazing intensity. Their population in grazing grassland were seriously declined by long – term overgrazing and became sparsely distributed and dwarfed remnant population, while their dominance were higher in forbidden grazing areas. Matthew and his colleagues (1995) pointed out that the single tiller density of *S. grandis* showed a single peak growth pattern with the increase of grazing intensity within a certain range. It is useful adaptability for *S. grandis* to make up for the loss of leaf area due to grazing. Therefore, when the grazing intensity is reduced or removed, *S. grandis* can increase its biomass in a short period of time. However, if the grazing intensity is out of the range of adaptability, the compensation characteristics disappear and the plant's aging rate and death rate accelerate.

An (2001) studied the responses of *S. grandis* in different seasons to grazing and found that there was a large difference in the net primary growth during different growth seasons. During early summer, moderate grazing could stimulate the growth of *S. grandis* tillers, and it is good for removing litter layers, improving micro-ecological environment, increasing light and temperature of the plant base, promoting the growth of tillers and thus, increasing the net primary growth. According to the equation, net accumulation of pasture = total growth − (dead weight + feed intake), Bircham and Hodgson (1983) pointed out that there are two reasons for the declining net primary growth of *S. grandis* in autumn under low grazing intensity (feed intake is relatively low): ① decrease in total plant growth; ② higher proportion of plant natural aging and death, whereby the effect of the latter was larger. Therefore, there is a suitable range of utilization rate for the *S. grandis* population. Rational use of *S. grandis* can reduce its rate of natural aging and death, increase its net primary growth and improve its utilization efficiency.

From the study of Li (1994), along with the increase of grazing intensity, the dominance of *L. chinensis* gradually reduced. Hence, it is a type of negative quantitative indicator plant. Besides, the dominance of *L. chinensis* increased with the increase of grazing intensity before it decreased. Therefore, moderate grazing can promote shooting and increase its dominance to a certain extent although it causes the species to become smaller in size. He also pointed out that *S. grandis* had similar results but would disappear in a high grazing intensity area and would be replaced by *S. sareptana*. The results of this study in this book showed that the dominance of *S. grandis* and *L. chinensis* was

Chapter 5　Discussion

greater in rotational grazing area than in continuous grazing area and the dominance of *S. grandis* (1.68%) was much lower than *S. sareptana* (14.13%) in continuous grazing areas. This phenomenon was also consistent with the above findings and theories.

The dominance of *C. duriuscula* and *A. frigida* was the highest in rotational grazing area, followed by continuous grazing area and lastly forbidden grazing area, of which the significant difference in the dominance of *C. duriuscula* was significantly greater in rotational grazing area than in both continuous grazing area and forbidden grazing area ($p < 0.05$). The dominance of *S. sareptana*, *C. acuminatum*, *C. squarrosa*, *A. polyrhizum*, *S. collina* and *A. tenuissimum* was the highest in continuous grazing area, among which the dominance of *S. sareptana*, *C. squarrosa*, *A. polyrhizum*, and *A. tenuissimum* was the second in rotational grazing area and the smallest in forbidden grazing area, and the dominance of other two species, *S. sareptana* and *C. squarrosa* was the second in forbidden grazing area and the smallest in rotational grazing area. The significant difference in the dominance of *S. sareptatna* and *C. squarrosa* was greater in continuous grazing area than in rotational grazing area and forbidden grazing area ($p < 0.05$) while the significant difference in the dominance of *A. tenuissimum* was greater in continuous grazing area than in rotational grazing area ($p < 0.05$). These results were consistent with those indicated by Li Yonghong (1994), who pointed out that *C. duriuscula*, *A. frigida*, *S. sareptana*, *C. acuminatum*, *C. squarrosa*, *A. polyrhizum*, *S. collina* and *A. tenuissimum* weree either positive qualitative indicator plants or positive quantitative indicator plants (Li, 1994). For example, *A. frigida* found in the study area, belongs to the lower layer's auxiliary species in non – degraded communities. During the degradation process, the absolute and relative amounts of aboveground biomass and density of *A. frigida* increased due to the decline of tall grass populations, to the point where it became the dominant population. Because *A. frigida* contains monoterpene, sesquiterpene, santonin substances (The Editors of *Chinese Herbs of China*, 1975) the plants are not preferred by those livestock during its growing season. The species, *A. frigida*, has two types of reproductive methods, namely wind spread seeds and vegetative propagation of stolon and adventitious roots. Therefore, two types of roots can be formed. First, the deep root system developed by the growing of seed can grow up to 70 cm into the soil layer and second, the shallow adventitious root system which can only grow 0 ~ 30 cm into the soil layer. With the degradation of community, *A. frigida* can expand its population through two effective reproductive methods and take up resources and space with its developed root system. This allows the species to develop into a dominant population in the degraded community, hence, a positive quantitative indicator to indicate strong grazing intensity.

The main plant types in the typical steppe will gradually change into *A. frigida* grassland under continuous heavy grazing. Therefore, *A. frigida* plays a very important role in the study of grazing succession. However, it does not mean that *A. frigida* grassland is the last stage of degradation due to grazing (Li, 1994). In extremely heavy grazing areas, *A. frigida* will disappear too and left only with *A. polyrhizum*, *A. tenuissimum* or *C. acuminatum*, *S. collina* and other annual plants, and even bare land. *A. polyrhizum* are densely clustered by numerous closely aggregated bulbs and are vegetative propagated by tillers. The species root system is rather swallowed and usually grows 0 – 20 cm into the soil uppermost layer only. Leaves are succulent with water storage tissues. If the plant water content is calculated with the formula, (fresh weight – dry weight) /dry weight, *A. polyrhizum* contains up to 300% of water while young leaves of *L. chinensis* contains up to 130 % of water only. The uppermost layer of soil where the root system of *A. polyrhizum* are concentrated, is usually the level with the greatest variation in soil water content. Therefore, it relies on its water storage feature to effectively resist the seasonal drought caused by uneven precipitation. This feature is not available in *C. squarrosa* though they share a similar root system, vegetative propagation method and other colonization means. In addition, *A. polyrhizum* population can form a more compact block in the community and also help to maintain the relative stability of their population density (Wang et al., 2011).

The degeneration of the grassland ecosystem due to the effects of grazing, has obvious adverse succession of the community. The dominance of *A. frigida* and *C. duriuscula* in rotational grazing area was the highest while the dominance of *C. acuminatum*, *A. polyrhizum*, *A. tenuissimum*, and *C. squarrosa* was the highest in continuous grazing area. From the structure of species dominance in the three comparison areas, the degradation degree in continuous grazing area was the highest, followed by rotational grazing area and lastly, forbidden grazing area. The *S. grandis* + *L. chinensis* grassland in forbidden grazing area would be transformed into a more grazing resistant *L. chinensis* + *C. duriuscula* + *A. frigida* + *C. squarrosa* grassland or *C. duriuscula* + *C. squarrosa* + *A. polyrhizum* + *C. acuminatum* + *S. collina* grassland.

It is worth stating that all above-mentioned indicator plants, especially quantitative indicator plants, are relative to a certain type of grassland or a certain area, hence indicating that the indicator plants are regional.

5.3 Effects of Different Grazing Systems on Species Diversity

Global change and human activities are affecting biodiversity at an unprecedented rate throughout the world (Grime, 1998; Valencia et al., 2015) and many issues associated with grazing have been addressed and studied in various ways (Li, 1999; Ding et al., 2012). Among them, the effect of grazing on species diversity is an important research direction. There have been a large number of studies on the relationship between the species diversity and grazing on the grassland's plant communities (Grime, 1973; Collins, 1987; Wang, 2005). The findings of this book indicated thatthere were no significant differences in species richness R, Shannon – Wiener index, Pielou's evenness index among the three different grazing system areas. These results indicated that different grazing systems in the cross – border areas of China and Mongolia have not yet had a significant effect on community species diversity. Besides, from the value of Simpson indexes being significantly greater in continuous grazing area and forbidden grazing area than in rotational grazing area, it can be seen that the effects of different grazing systems on community species diversity were shifting towards the direction of significant difference ($p < 0.05$) (Figure 4 – 5 to Figure 4 – 8).

The results showed that the species richness, R, reduces from continuous grazing area > rotational grazing area > forbidden grazing area (Figure 4 – 5), because of perennial grasses are preferable and largely grazed by livestock, leaving more space and resources for weeds and ephemeral plants to survive (Mcintyre et al., 2003). Meanwhile, grazing causes dwarfing of plant and plasticity of plant morphology, resulting in a greater species richness index in continuous grazing area and rotational grazing area than in forbidden grazing area. This was in line with the findings of many previous studies (Oba et al., 2001; Altesor et al., 2005). Some scholars also believed that forbidden grazing system could increase the species diversity and excessive disturbance would lead to the disappearance of some species, hence, reducing the species diversity (Sternberg et al., 2000; Akiyama & Kawamura, 2007; Firincioglu et al., 2007).

For example, strong grazing intensity will reduce or remove the palatable pasture while under a forbidden grazing system, the palatable pasture will increase, hence, increasing the species richness and diversity (Marcelo et al., 2000). Livestock grazing dynamically regulates the species richness of the community by altering the rate of colonization and extinction of local plant species. When the extinction rate is lower than the colonization rate, it does not reduce the local species richness. On the contrary, when the extinction rate is higher than the colonization rate, it will not only reduce the local species richness but also may lead to the extinction of the entire community (Han & Ritchie, 1998; Olff & Ritchie 1998; Wang et al., 2001). According to the data of this study, continuous grazing area with the highest grazing intensity when compared to the other two systems had not yet led to the disappearance of many species but had a slightly higher species diversity than in both forbidden grazing area and rotational grazing area. Although there was no significant difference in the stocking rate between rotational grazing area and continuous grazing area, the grazing intensity in rotational grazing area was lower than in continuous grazing area while the species and communities' average height, total coverage and aboveground biomass were far higher than those in continuous grazing area. *A. frigida* and *C. duriuscula* were at obvious advantage, which to some extent, they suppressed the usual competitive growth among the inferior species such as *A. frigida*, *C. duriuscula* and *C. squarrosa*, resulting in the species richness in rotational grazing area was lower than in continuous grazing area. In addition, the study area of 1,400 km^2 with similar ecological factors in the cross – border areas of China and Mongolia, which has no significant difference in grazing intensity, is also an important reason for no significant difference in species richness among the three different grazing system areas.

From the result of Pielou's evenness index which expressed the distribution of the number of individual species, the index reduced from forbidden grazing area > continuous grazing area > rotational grazing area (Figure 4 – 8). The main reason may be that the forbidden grazing area is not affected by grazing, hence, vegetation distribution is relatively even and the relationship between species is stable. The disequilibrium among species caused by grazing disturbance of typical grassland vegetation communities at higher grazing intensity is much greater than the disequilibrium among individual species under natural condition (Liu et al., 2013). The difference between continuous grazing and rotational grazing is mainly due to continuous grazing area has high grazing intensity, serious grassland degradation, low option of feed which force the livestock to feed on less palatable plants in great amount. Therefore, the species in continuous grazing areas cannot become dominant species, resulting in a relatively even distribution of the number of individual species in continuous grazing areas.

The trend of change for Shannon – Wiener index and Simpson index was basically similar, where continuous grazing area > forbidden grazing area > rotational grazing area (Figure 4 – 6 and Figure 4 – 7). The difference of Shannon – Wiener index for the three comparison areas was very small as the difference between the largest index value of continuous grazing area and the smallest index value of rotational grazing area was less than 0.1. This is closely related to the above selected experimental site.

In grassland ecosystem with better nutrient conditions, some research findings support the hypothesis of intermediate disturbance is mainly because: Under the condition without grazing disturbance, tall plant species have absolute advantage in the community and reduce the amount of light transmitted in the lower layers of the community, which restrains the increase of species diversity; moderate grazing disturbance reduces the competitive advantage of tall plant species, allowing the species of lower layer to increase significantly and thus, presenting the distribution pattern whereby both tall plant species and dwarf plant species coexist, hence, increases the species diversity; under high grazing intensity condition, the communities have only a small number of grazing resistant species, hence, significantly reduces the species diversity (Connell, 1978; Milchunas et al., 2014). However, the typical steppe in Inner Mongolia is an ecosystem limited by both water and nitrogen, hence, the species are mainly competing for underground resources such as water and nutrients. Owing to the limitation of water and nutrients, the community coverage is relatively low and the light competition among species is weak. Grazing further reduces the community coverage and inhibits the growth of non-grazing resistant species. Therefore, the typical steppe in Inner Mongolia demonstrates a pattern whereby the species diversity decreases with the increase of grazing intensity (Bai et al., 2004). Moderate grazing disturbance has obvious ecological thresholds for different grassland community types and different grazing intensity. In the wetter grassland of Mongolian steppe, moderate disturbance happens during low or intermediate grazing intensity which is similar to the findings of Yang (2001) and Sasaki (2008). The more arid typical steppe is more resistant to grazing and moderate disturbance happens only during intermediate or strong grazing intensity while the arid typical steppe does not support the theory of moderate disturbance at all (Yang & Han, 2001). From the results of this study, the Shannon – Wiener index was greater in continuous grazing area than in both forbidden grazing area and rotational grazing area, which further indicated that moderate disturbance of typical steppe may occur when the grazing intensity is strong. Compared to continuous grazing, rotational grazing has the effect of reducing grazing pressure. The community degraded from the absolute dominance of *S. grandis* and *L. chinensis* in forbidden grazing area, to the dominance of *A. frigida*, *C. duriuscula* etc, in rotational grazing area and the species diversity was lower than that in continuous grazing area. However, some studies also suggested that the grazing conditions will change the intensity of competition among grassland species, resulting in the exclusion of some inferior species and leading to a decrease in the community's species diversity (Sun et al., 2013).

Besides, the presence of a large number of rare species is an important indicator of high biodiversity (Godron et al., 1971; Mouillot et al., 2013). This is another reason why the species diversity in both forbidden grazing areas and rotational grazing areas in this study was low. Because the number of auxiliary species and rare species is small, their distribution range is limited and most of the rare species have good palatability, they easily become the feed of livestock under rotational grazing, resulting in their decreasing number or even extinction. Examples are *Agropyron cristatum*, *Heteropappus altaicus*, *Allium tenuissimum* and *Allium ramosum* that are with larger leaves, higher water content and better palatability (Zhang, 2011). In contrast, previous studies which calculated the diversity index based on biomass data suggested that dominant species have higher biomass and thus, play a leading role in ecosystem functioning while rare species are secondary and have limited impacts on ecosystems (Kraft et al., 2011; Mouillot et al., 2013). The study of this book calculated the species dominance by relative height, relative coverage, relative density and relative frequency and finally obtained the Shannon – Wiener index without taking the biomass data of each species to calculate the diversity index. During the actual survey, continuous grazing area had relatively high degree of degradation, large degree of fragmentation of tussock perennial grasses and other plant species and higher calculated frequency and number of individual species than in rotational grazing area and forbidden grazing area, which might affect the species dominance, and thus the species diversity in continuous grazing area was high.

The Simpson index of forbidden grazing area and continuous grazing area were basi-

Chapter 5 Discussion

cally equal and the significant difference was greater than that in rotational grazing area (p <0.05). This result seemed to be inconsistent with the Moderate Disturbance Hypothesis. In fact, the feeding of livestock is a very complex ecological process. Livestock do not only affect plant species diversity, plants also react with appropriate countermeasure to limit the behavior of livestock (Wang et al., 2010). When an individual of a plant coexists with a species with better palatability to form a neighbor relationship, the animal is attracted by the plant with better palatability, thereby reducing the need to feed on that particular species (Danell et al., 1993). In a typical steppe with relatively few species, *S. grandis* and *L. chinensis* are the two species at advantage. According to the dominance data in rotational grazing area, less palatable species such as *C. duriuscula* grew steadily above *L. chinensis* and below it, were the plants with least palatability such as *A. frigida* and *C. acuminatum*. Although the degrading phenomenon was obvious in the rotational grazing area, the dominance of *S. grandis* and *L. chinensis* was higher when compared to continuous grazing area. During foraging, sheep prefer species with higher palatability such as *S. grandis* and *L. chinensis*, resulting in a gradual decrease in the dominance of *S. grandis* and *L. chinensis* and a gradual increase in the dominance of *C. duriuscula* and *A. frigida*. Therefore, the difference in the dominance among species was large but the Simpson dominance index was relatively low, which might be related to the fact that Simpson dominance index of rotational grazing area was smaller than that in forbidden grazing area. The reason for the difference in Simpson dominance index between continuous grazing area and rotational grazing area are basically similar to the difference in Pielou evenness index between continuous grazing area and rotational grazing area, ie. the grazing intensity in the continuous grazing area was greater than in the rotational grazing area.

5.4 Effects of Different Grazing Systems on Plant Functional Groups

Functional groups serve as bridges to connect environment, individual plants and ecosystem structures, processes and functions together (Kleyer, 2002; Cornelissen et al., 2003) and thus, the use of functional groups for related research has become effective and convenient means.

Plant functional groups respond similarly as a group to external disturbances and environmental impacts, and their major plant groups have similar ecological change processes (Lavorel & Garnier, 2002). Selective feeding of livestock increases the heterogeneity of habitats and thus changes the composition, structure and diversity of species communities in grassland, which affect the structure and function of the entire ecosystem (Yang et al., 2001). Grassland productivity and community structure are largely affected by the diversity and composition of plant functional groups (Tilman, 2001). Precipitation, environmental changes and grazing help to increase the diversity of functional groups and the coexistence of species (Lorenzo et al., 2012).

Table 4 – 3 showed that the dominance of xerophytes and intermediate mesophytes reduced from continuous grazing area > forbidden grazing area > rotational grazing area, and the significant difference in the dominance of xerophytes was greater in continuous grazing area and forbidden grazing area than in rotational grazing area but there was no significant difference between continuous grazing area and forbidden grazing area. There was no significant difference in the dominance of intermediate mesophytes among three different grazing system areas.

The dominance of intermediate xerophytes and mesopytes reduced from rotational grazing area > forbidden grazing area > continuous grazing area, and the significant difference of the dominance of intermediate xerophytes was greater in rotational grazing area than in forbidden grazing area and continuous grazing area but there was no significant difference between continuous grazing area and forbidden grazing area. There was no significant difference in the dominance of mesophytes among three different grazing system areas. The above results indicated that the water – based functional groups have significant differences in areas with different grazing systems. The trend of aridification is significantly greater in continuous grazing areas and forbidden grazing areas than in rotational grazing areas. The main reason could be because the forbidden grazing area completely excludes grazing disturbance while continuous grazing area has too frequent and strong disturbances. Both conditions are not conducive to soil moisture retention in grassland but the condition of rotational grazing area is conducive to soil moisture retention in grassland. Based on the survey data, the total aboveground biomass, total coverage and average height were greater in the rotational grazing area than those in continuous grazing area (Figure 4 – 2 and Figure 4 – 3). Vegetation biomass and coverage can prevent or reduce the direct exposure of soil to the sun, hence effectively reduce the water moisture in soil from evaporation (Yang, 2005).

One study found that appropriate grazing shows greater soil respiration rate during the dry season and facilitates the soil of grassland to absorb and utilize the rainwater (Hou et al., 2011). Mongolian grasslands are rotationally grazed according to the pasture growth and precipitation conditions. Such moderate grazing inhibits the plants from growing too tall and promotes the growth of tillers so that the number of individuals with the same genetic units increased (Wang & Wang, 1999). It also improves precipitation utilization, at the same time reduces the direct scouring effect of rainwater on plant root soil and prevents soil erosion (Li et al., 2003). If the grazing disturbances are too frequent and intense, the compensatory characteristics of many species will disappear, which accelerate the individual rate of aging and death, rapidly decrease the aboveground biomass and vegetation cover and eventually expose a large area of topsoil (Zhao, 1999). After the destruction of surface vegetation, in addition to increasing the air mobility of the surface and enhancing evaporation conditions, the consumption of solar energy by photosynthesis under the direct irradiation of equal sunlight decreases, hence the solar energy received by the soil increases. For this reason, the soil temperature and evaporation from the surface increase, and the soil water content decreases (Tong et al., 2000; Lin, 2007). Besides, high trample frequency of livestock which shrink the soil pores, reduce soil respiration rate and infiltration rate, and no interception of vegetation layer have made the soil layer prone to soil erosion and significantly reduced precipitation utilization (Gan et al., 2012). These may lead to a more severe tendency of aridification of the surface soil layer and the entire ecosystem, promoting a significant increase in the dominance of grazing – tolerant and drought – tolerant species (Su et al., 2005).

Due to the lack of soil loosening and crushing caused by the trampling of livestock in the forbidden grazing area, the ground surface is hardened and the infiltration capacity of precipitation is reduced, making it prone to soil erosion. Besides, lacking livestock manure input causes nutrient deficiency that eventually leads to soil aridity and barrenness (Bao et al., 2009). Without the disturbances from livestock in forbidden grazing areas, the protein content of pasture decreases with maturity while the fiber content increases with maturity. This will affect the compensatory growth of pasture, reducing the ability to directly absorb precipitation and affecting the degree of precipitation utilization. Therefore, the dominance of species resistant to grazing and drought is higher in forbidden grazing areas (Xi, 2008). However, it has also been pointed out that soil nutrients and moisture are influenced by multiple factors such as the natural conditions of the grassland, the duration of grazing exclusion, soil texture and mechanical composition (Su et al., 2005).

The analysis of the dominance of community life-form functional groups in different grazing system areas revealed that (Table 4-4) the dominance of perennial grass and shrubs and sub-shrubs reduced from forbidden grazing area > continuous grazing area > rotational grazing area, of which the significant difference in the dominance of perennial grass was greater in forbidden grazing area and continuous grazing area than that in rotational grazing area but there was no significant difference between forbidden grazing area and continuous grazing area and no significant difference in the dominance of shrubs and sub-shrubs among the three different grazing system areas. The dominance of perennial weed and annual grass reduced from rotational grazing area > continuous grazing area > forbidden grazing area, of which the significant difference in the dominance of perennial weed was greater in rotational grazing area than in both forbidden grazing area and continuous grazing area but there was no significant difference between forbidden grazing area and continuous grazing area. There was no significant difference in the dominance of annual grass among the three different grazing system areas. The main reasons could be due to the grazing effect of livestock in rotational grazing area inhibit the dominant species plants such as *S. grandis* and *L. chinensis* from growing taller, hence, facilitating the dominance of perennial weed such as *A. frigida*, *C. duriuscula*, *A. polyrhizum*, *H. altaicus* and *O. mocrphylla*. On the other hand, rotational grazing reduces grazing intensity, which to a certain extent facilitates the improvement of soil water retention capacity of vegetation cover, and the dominance of some annual grasses such as *S. collina*, *C.*

acuminatum, *A. annua*, *D. micranthus* etc, increases significantly.

Li (1993) believed that the species diversity of *L. chinensis* grassland and *S. grandis* grassland on grazing intensity depends on interspecific competitive exclusion of communities and grazing inhibition or promotion of different plant growth. Therefore, strong community interspecific competition in forbidden grazing areas inhibits the growth of most plant species, allowing *S. grandis*, *L. chinensis* and other perennial grasses to take an absolute dominance position. In all three different grazing system areas, the dominance of perennial grass was the highest while the dominance of perennial weeds was lower because they were being suppressed. However, shrubs such as *C. stenophylla*, *C. microphylla* etc and sub – shrubs such as *E. sinica* etc are large in size, have a deep root system and are strongly resistant to drought, hence, are less affected. Besides, because forbidden grazing prevented livestock feeding, the dominance of shrubs and sub – shrubs species in forbidden grazing areas was the highest. Liu (2004, 2006) believed that the longer the grassland is enclosed, the more obvious the vegetation turning into shrubs which reduce edibility and even cause adverse succession and degradation.

According to the study of Zuo and others in 2009, under heavy grazing pressure, grassland degradation occurs and the community structure tends to be simple. In all stages of degradation, the functional groups of drought – tolerant and grazing – tolerant perennial grass maintain a higher dominant position, hence playing an important role in maintaining the ecological function of the community. At the same time, shrubs and sub – shrubs in grassland is another important manifestation of grassland degradation. These supported the findings of this book that the dominance of perennial grasses and shrubs and sub – shrubs in the continuous grazing area was second only to the forbidden grazing area. The trend of aridification was more severe in continuous grazing areas and forbidden grazing areas than in rotational grazing areas, and the dominance of annual grass that preferred higher humidity habitats was lower.

In addition, the survey found that the phenomenon whereby *S. grandis* was being replaced by *S. sareptana* in continuous grazing areas was very obvious (Table 4-2). The continuous grazing system in Inner Mongolia has caused a large area of grassland being fenced into smaller areas and a loss of accessibility. Therefore, the number of horses and camels which require a large area of grassland for feeding and long-distance walking as well as also prefer to feed on grasses of genus Stipa was much smaller than those in Mongolia (Table 5-1). Grassland with grasses of genus Stipa is usually more valuable in spring and early summer prior to the maturity season of caryopsis because mature caryopsis has hard and sharp branches, which often puncturing the sheep' mouth and skin, affecting their health or mix with wool and influence wool quality. Nonetheless, grassland with mature caryopsis of Stipa does not affect larger size livestock such as camels, horses and cattle from feeding. Rotational grazing can more effectively and flexibly adjust the use of different pastures, including the grassland with grasses of genus Stipa, by different types of livestock in different seasons. Even though the quantity of camels, horses and other larger size livestock is rather few in Inner Mongolia Autonomous Region, they are still affecting the Stipa grasses and shrubs and sub-shrubs to a certain extent because of their feeding behavior, resulting in the higher dominance of *S. sareptana* and shrubs and sub-shrubs species after the degradation of *S. grandis* in continuous grazing area.

蒙古草原放牧制度效益评价研究
Evaluation Research on the Benefits of Grazing System in the Mongolian Grassland

Table 5 – 1   The number of livestock in Naren and Naran Soum in 2016

	Naren	Naran
Camels	62 (0.02%)	677 (0.31%)
Horses	1,988 (0.61%)	18,854 (8.51%)
Cattles	17,415 (5.33%)	11,348 (5.12%)
Sheep	259,289 (79.39%)	108,902 (49.13%)
Goats	47,857 (14.65%)	81,859 (36.93%)
Total	326,611	221,640

Note: Data in parentheses are the proportion of the number of each livestock to the total number of livestock.

5.5   Effects of Different Grazing Systems on NDVI

NDVI is very sensitive towards the change in ecosystem (Sternberg et al., 2011). Therefore, when all abiotic factors such as weather, topography and soil as well as grazing intensity are generally the same in the area, the grazing systems could be the main cause which affect the grassland ecosystem (Zhang et al., 2007; Wang et al., 2013).

Naran Soum in Mongolia introduced private ownership of livestock from 1990 however a rotational grazing system is still the main method of livestock management. Inner Mongolia has had a private ownership system since 1978 but the implementation of the Pasture Houshold Contract Responsibility System in 1990 has gradually replaced the use of rotational grazing system to continuous grazing system at Naren Soum.

From 1989 to 2016, for a period of 28 years, the average stocking rate of pastoralists at the sampling location in Naran Soum of Mongolia was 42 sheep/ km^2 and 50 sheep/ km^2 at Naren Soum of Inner Mongolia, with no significant difference ($p > 0.05$) (Table

2 – 6).

The five periods NDVI results showed that the NDVI values in the rotational and continuous rotational grazing areas differed over time, i. e., grassland degradation in the continuous grazing area was greater (or lower NDVI) than in the rotational grazing area (Figure 4 – 9 to Figure 4 – 13).

After James Anderson proposed the theory of rotational grazing at the end of the 18th century, many scholars conducted the study at various places and supported his viewpoint, ie. rotational grazing can increase grass production and improve grassland utilization (Derner et al., 1994; Jacobo et al., 2006). One study also explained that rotational grazing can facilitate grassland recovery, increase vegetation coverage and pasture quality (Savory & Stanley, 1980). In particular, choosing a suitable grazing system based on the differences in topography conditions can improve grassland utilization, prevent degradation and be beneficial to livestock production (Hao et al., 2013). However, Derek W. Bailey proposed that in arid and semi – arid shrub lands, timely adjustments of grazing intensity maintains and improves ecological health at regional and landscape scales better than rotational grazing and forbidden grazing (Bailey & Brown, 2011). Martin and his colleagues explained that rotational grazing can facilitate grassland recovery of unhealthy grassland but such effects is minimum when the grassland is healthy (Severson, 1988); Heitschmidt and his team (1987) selected cows in their experiment in Texas, and concluded that the impacts of rotational grazing and continuous grazing on the environment were basically similar and the differences are mainly caused by the difference in grazing intensity. In addition, Heitschmidt (1982) also found that different grazing seasons and grazing systems affect vegetation differently.

The results of NDVI dynamics in the forbidden grazing area and grazing areas showed that forbidden grazing has a protective effect on above – ground biomass and cover for the same reasons as the difference in the four general characteristics of plants (average height, total coverage, total individual density and total aboveground biomass) between the forbidden grazing area and grazing areas mentioned above. There was no significant difference in the NDVI values of year 2005 between forbidden grazing area and rotational grazing area (Figure 4 – 11) but the NDVI value in rotational grazing area was larger than that in forbidden grazing area in 2011 (Figure 4 – 12). These data indicated that the NDVI values in grazing areas can be larger than in the forbidden grazing area under certain conditions. For instance, because the precipitation of Naran Soum in 2005 (90 mm) was lower than the average rainfall (216.80 mm), NDVI values of the study area were relatively lower, and there was no significant difference in NDVI values between rotational grazing area and forbidden grazing area. However, because the precipitation of Naran Soum in 2011 (217.56 mm) was higher than the average rainfall, NDVI values of the study area were relatively higher, of which the NDVI value in rotational grazing area was higher than that in forbidden grazing area though there was no significant difference between them. This result is consistent with the view that moderate grazing can increase species richness and biomass of grassland communities as indicated by some researchers. They also pointed out that the species diversity and functional diversity are greatest during moderate grazing and thus, it can ensure sustainable use of grassland as it increases complexity and stability of plant community structure (Lavorel et al., 1997; Ruifrok et al., 2014). In addition, Collins (1987) and Loeser (2007) showed that long periods of forbidden grazing were not conducive to maintaining high community diversity and productivity. Combining the above findings with the results of this study, which showed that the total density and abundance of individuals in the forbidden grazing area were lower than those in the rotational grazing and continuous grazing areas, it was found that further examination is needed to determine the superiority of the forbidden grazing system and other two grazing systems.

5.6 Effects of Different Grazing Systems on Plant Community Stability

The concept of stability comes from system cybernetic and often refers to the convergence of system deviation after the system is subjected to external disturbances, or the value of the system deviates from the equilibrium position. After introducing the concept of stability into ecosystem research, it has aroused widespread controversy in the field of modern ecology. New hypotheses and views are constantly being introduced, denied and amended. Although there are still many disagreements to the current understanding of ecosystem stability, people have initially reached consensus on the stability of the ecosystem after a long-term development. The stability of the ecosystem includes: ①the ability of the ecosystem to maintain its status quo, that is, the ability to resist interferences; ②the ability of the ecosystem to return to its original state after being disrupted, that is, the ability to recover after disturbance (Ma, 2002).

Even though ecologists have proposed some methods to measure the stability of ecosystems (Goodman & Daniel, 1975), all these methods have different degrees of shortcomings and cannot be fully and effectively applied to the evaluation of actual ecosystem stability. Therefore, other researchers put forward their own stability evaluation indicators for different ecosystems (Xiao et al., 2003; Zhang, 2006).

This book studies ecosystem stability from a community perspective. The stability of the plant community is a very complex issue as it includes the community composition, function and all interference factors (Wang et al., 2006). Though many scholars have done a great deal of work, many issues are still deserving for more in - depth studies. The number of species in a community and the number of individuals in a species, to a certain extent, reflect the characteristics of the community, the stage of development and the degree of stability of the community. In order to comprehensively reflect the actual situation of the community stability of typical grassland in cross - border areas of China and Mongolia, this study revised the M. Godron's stability test in measuring community stability and calculated both frequency - based and coverage - based M. Godron's community stability.

The calculated results of frequency - based M. Godron's community stability test (Table 4 - 5 and Figure 4 - 14) showed that the stability reduced from continuous grazing area > rotational grazing area > forbidden grazing area. Among the three comparing areas, the area with high species richness was relatively stable while the area with low species richness was unstable. Such results are consistent with the experiment of Liu Jingling and her team (2006) on grassland plant communities in the central - eastern part of Inner Mongolia.

The calculated result of coverage - based M. Godron's community stability test (Table 4 - 6 and Figure 4 - 15) showed that the stability reduced from rotational grazing area > continuous grazing area > forbidden grazing area. The reasons may be that although the coverage in forbidden grazing areas was greater than in rotational grazing and continuous grazing areas (Figure 4 - 2) and there is a large difference between the coverage of each species (Figure 5 - 1). For example, the coverage of species such as *S. grandis*, *L. chinensis*, and *K. prostrata* in forbidden grazing areas were 25.71%, 13.71% and 12.86% respectively, of which their coverage was at absolute advantage.

Chapter 5　Discussion

The coverage of other species was less than 6.44%. The species richness in both rotational grazing area and continuous grazing area were greater than in forbidden grazing and when compared to forbidden grazing area, the coverage among species was more even. For example in the rotational grazing area, the dominance of *C. duriuscula*, *L. chinensis* and *A. frigida* were 23.37%, 16.20% and 11.48% respectively. In continuous grazing areas, other than *S. sareptana*, 15%, and *C. squarrosa*, 7.26%, were more prominent, the coverage of other species was more even. The difference between continuous grazing and rotational grazing may be due to the coverage in continuous grazing areas being significantly lower than rotational grazing areas (Figure 4 – 2), and greater in variation (Figure 5 – 1). For instance, the average total coverage in rotational grazing area was 64.30%, with standard deviation of 9.82 while the average total coverage in continuous grazing area was 56.52%, with standard deviation of 17.42. This result is similar to the result of Bai and his colleagues (2000).

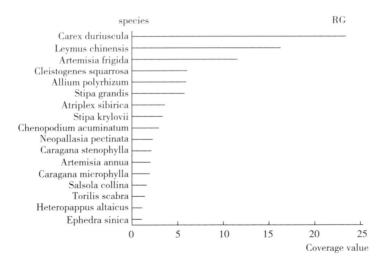

Figure 5 – 1　The coverage value in different types of grazing system areas (Coverage value >1%)

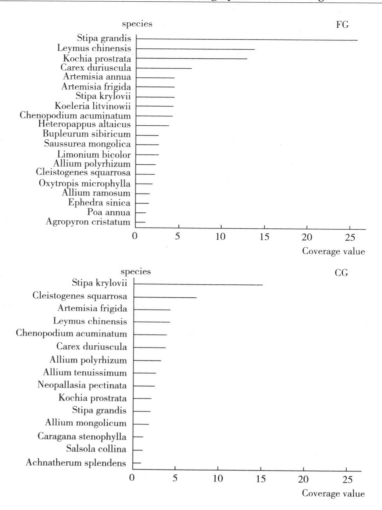

Figure 5–1 The coverage value in different types of grazing system areas (Coverage value >1%) (Continued)

Pimm (1984) pointed out that there is no simple correlation between species ecological diversity and ecological stability. Yang et al. (2006) also found out that the correlation between species ecological diversity and ecological stability is not significant. There are two reasons to explain this situation (Boucot, 1985; Yang et al., 2006):

(1) The changes in dominance of foundation species. When the foundation species are late successional plant species with relatively large dominance, there is good consistency in species diversity and stability. Meanwhile, when the foundation species are early succession plant species with absolute dominance, there is also a good consistency in species diversity and community stability. In contrast, when plant species are relatively more dominant in the early successional stage and less dominant in the late successional stage, community species diversity and community stability are inconsistent.

(2) Grazing disturbance. There is no simple linear relationship between species diversity and community stability but rather there is a significant uncertainty. Grazing can increase and decrease species diversity and can lead to changes in the stability of the community.

Therefore, the study of stability should emphasize on the basic and impact factors of diversity and stability. Combining species richness, species characteristics, community structure and interference factors to study the stability, thereby promoting the studies on the stability mechanism of the natural grassland.

Chapter 6   Conclusion

(1) The plant community'average height, total coverage and total aboveground biomass reduced from forbidden grazing > rotational grazing > continuous grazing. There was no significant difference in the average height and total aboveground biomass among the three different types of grazing system areas ($p < 0.05$). The significant difference of the total coverage was greater in forbidden grazing area and rotational grazing area than in continuous grazing area ($p < 0.05$) but there was no significant difference between rotational grazing area and forbidden grazing area. It can clearly be seen that grazing systems significantly change the basic characteristics of vegetation communities in the study area. The main reason may be the behavior of livestock such as feeding and trampling that reduce the leaf surface area, decrease the plants' ability of photosynthesis as well as change the number and structure of pasture, which in turn affects the basic characteristics of plant communities.

(2) The responses of 10 species with dominance greater than 3% to different grazing systems were analyzed, and it was found that different grazing systems produced significant differences in the effects on different species. Though the grassland degradation and adverse succession were very obvious in the study area, the typical steppe was still

Chapter 6 Conclusion

maintaining the perennial grass – dominated plant communities that are resistant to grazing and drought. *S. grandis* and *L. chinensis* were at absolute dominance in the forbidden grazing area but their dominance decreased in the rotational grazing area where *A. frigida* and *C. duriuscula* were at absolute dominance. After *S. grandis* degenerated in the continuous grazing area, *S. sareptana* was the dominant species. Besides, the dominance of other species that are resistant to grazing and drought such as *C. acuminatum*, *A. polyrhizum*, *A. tenuissimum* and *C. squarrosa* were obviously increasing in continuous grazing areas.

(3) There was no significant difference for species richness R, Shannon – Wiener index, and Pielou evenness index among the three different types of grazing system areas. This indicated that different grazing systems in the cross – border areas of China and Mongolia have not yet had a significant effect on community species diversity. Or as evidenced by the Simpson index being significantly greater in continuous grazing area and forbidden grazing area than in rotational grazing area, this effect is developing towards significant differences. Because grasses preferred by livestock are heavily grazed, weeds and ephemeral plants have more living space and resources to survive. Meanwhile, the dwarfing of plants due to grazing and the plasticity of plant morphology led to a greater species diversity index in continuous grazing area and rotational grazing area than forbidden grazing area. The Shannon – Wiener index was greater in continuous grazing area than in forbidden grazing area and rotational grazing area. It is because the moderate disturbance of typical grassland may occur at the stage of heavy grazing rate. Compared to continuous grazing, rotational grazing has the effect of reducing grazing pressure. The communities degraded from the absolute dominance of *S. grandis* and *L. chinensis* in forbidden grazing area to the dominance of *A. frigida*, *C. duriuscula* etc in the rotational grazing area, where the species diversity is not as good as in continuous grazing area.

(4) In areas with different grazing systems, the dominance of water – based functional groups and life – form functional groups are significantly different. The significant difference of xerophyte was greater in continuous grazing area and forbidden grazing area than in rotational grazing area while the significant difference in the dominance of intermediate xerophyte was greater in rotational grazing area than in forbidden grazing area and continuous grazing area. The significant difference in the dominance of perennial grass was greater in forbidden grazing area and continuous grazing area than in rotational grazing area but there was no significant difference between forbidden grazing area and continuous grazing area. The significant difference in the dominance of perennial weeds was greater in rotational grazing area than in forbidden grazing area and continuous grazing area but there was no significant difference between forbidden grazing area and continuous grazing area. It can be seen that the tendency of aridification in continuous grazing area and forbidden grazing area was significantly higher than in rotational grazing area, and thus the dominance of drought – tolerant and grazing – tolerant perennial grasses, shrubs and sub – shrubs increases. The main reason may be due to a certain range of grazing intensity is more conducive for grassland to maintain its soil moisture and improve the vegetation community.

(5) Among the five periods NDVI data, the NDVI values in 1989, 2005, 2011 and 2016 were higher in rotational grazing area than in continuous grazing area, and among them, the NDVI values in 2011 and 2016 were significant higher in rotational grazing area than in continuous grazing area. The NDVI dynamics in forbidden grazing area and grazing areas indicated that forbidden grazing plays a protective role on aboveground biomass and coverage, for the same reasons as those mentioned above for the differences in the basic characteristics of the communities.

(6) The calculated result for frequency – based M. Godron's community stability

Chapter 6 Conclusion

test showed that the stability reduced from continuous grazing area > rotational grazing area > forbidden grazing area, which manifested that the area with greater species richness was relatively stable while the area with lower species richness was unstable. The calculated result for coverage – based M. Godron's community stability showed that the stability reduced from rotational grazing area > continuous grazing area > forbidden grazing area. The total coverage may be closely related to species coverage but the stability of plant communities is a very complex issue as it includes the community's composition, functions and all interference factors.

Comprehensive analyses showed that under the same natural conditions and grazing intensity, the results of the current quadrat survey and dynamic NDVI survey demonstrated that forbidden grazing area's average height, total aboveground biomass and NDVI values of most years were significantly greater than in grazing areas; rotational grazing area's average height, total coverage, total individual density, total aboveground biomass and NDVI values of most years were higher than in continuous grazing area but the cumulative growth height and cumulative biomass etc throughout the growing period of forbidden grazing area and grazing areas require plant compensatory growth experiments to distinguish the dominance of forbidden grazing and grazing; community species diversity had yet to be significantly different; because of improper grazing, the degradation and adverse succession of grassland communities were very obvious, from the *S. grandis* + *L. chinensis* grassland in forbidden grazing area to a more grazing resistant *L. chinensis* + *C. duriuscula* + *A. frigida* + *C. squarrosa* grassland or *C. duriuscula* + *C. squarrosa* + *A. polyrhizum* + *C. acuminatum* + *S. collina* grassland; the significant difference in functional groups was significant, the trend of aridity was significantly greater in continuous grazing area and forbidden grazing areas than in rotational grazing area, and thus the dominance of drought – tolerant and grazing tolerant perennial grass and shrubs and sub – shrubs increased.

These indicated that under the same natural conditions and grazing intensity, the effects of different grazing systems on typical grassland plant communities were significant and to a certain extent, the rotational grazing system was better than continuous grazing system. However, due to the differences in vegetation type, grazing intensity, grazing period, livestock type as well as differences in their proportions and other environmental conditions, the impact of grazing systems on the plant community is rather complicated. Therefore, in – depth studies using various research methods such as correlation analysis and principal component analysis, taking into account spatial differences, community structure and interspecific relationships are needed in near future.

References

[1] Akiyama T., K. Kawamura. Grassland Degradation in China: Methods of Monitoring, Management and Restoration [J]. Japanese Society of Grassland Science, 2007, 53 (1): 1 – 17.

[2] Allen D. M. Planned Beef Production [J]. Outlook on Agriculture, 1985, 14 (1): 41 – 47.

[3] Altesor A., M. Oesterheld, E. Leoni, et al. Effect of Grazing on Community Structure and Productivity of a Uruguayan Grassland [J]. Plant Ecology, 2005, 179 (1): 83 – 91.

[4] Anderson Briskev. Tiller Dispersion in Populations of the Bunchgrass Schizachyrium Scoparium: Implications for Herbivory Tolerance [J]. Oikos, 1990, 59 (1): 50 – 56.

[5] Angerer J., G. Han, I. Fujisaki, et al. Climate Change and Ecosystems of Asia with Emphasis on Inner Mongolia and Mongolia [J]. Rangelands, 2008, 30 (3): 46 – 51.

[6] Armour C., D. Duff, W. Elmore. The Effects of Livestock Grazing on Western Riparian and Stream Ecosystem [J]. Fisheries, 1994, 19 (9): 9 – 12.

[7] Austrheim G. , O. Eriksson. Plant Species Diversity and Grazing in the Scandinavian Mountains—Patterns and Processes at Different Spatial Scales [J]. Ecography, 2008, 24 (6): 683 – 695.

[8] Bai Y. , X. Han, J. Wu, et al. Ecosystem Stability and Compensatory Effects in the Inner Mongolia Grassland [J]. Nature, 2004, 431 (7005): 181 – 184.

[9] Bailey D. W. , J. R. Brown. Rotational Grazing Systems and Livestock Grazing Behavior in Shrub – Dominated Semi – Arid and Arid Rangelands [J]. Rangeland Ecology & Management, 2011, 64 (1): 1 – 9.

[10] Baker M. J. Grasslands for Our World [M]. Wellington: SIR Publishing, 1993

[11] Bircham J. S. , J. Hodgson. The Influence of Sward Condition on Rates of Herbage Growth and Senescence in Mixed Swards under Continuous Stocking Management [J]. Grass & Forage Science, 1983, 38 (4): 323 – 331.

[12] Bisigato A. J. , B. B. Mónicar, J. O. Ares, et al. Effect of Razing on Plant Patterns in Arid Ecosystems of Patagonian Monte [J]. Ecography, 2005, 28 (5): 561 – 572.

[13] Boucot A. J. The Complexity and Stability of Ecosystems [J]. Nature, 1985, 315 (6021): 635 – 636.

[14] Burke I. C. , C. M. Yonker, W. J. Parton, et al. Texture, Climate, and Cultivation Effects on Soil Organic Matter Content in U. S. Grassland Soils [J]. Soil Science Society of America Journal, 1989, 53 (3): 800 – 805.

[15] Butler Briskej. Density – Dependent Regulation of Ramet Populations Within the Bunchgrass Schizachyrium Scoparium: Interclonal Versus Intraclonal Interference [J]. Journal of Ecology, 1989, 77 (4): 963 – 974.

[16] Carlson T. N. , D. A. Ripley. On the Relation Between NDVI, Fractional

Vegetation Cover, and Leaf Area Index [J]. Remote Sensing of Environment, 1997, 62 (3): 241-252.

[17] Carrera A. L., M. B. Bertiller, C. Larreguy. Leaf Litterfall, Fine – Root Production, and Decomposition in Shrublands with Different Canopy Structure Induced by Grazing in the Patagonian Monte, Argentina [J]. Plant & Soil, 2008, 311 (1-2): 39-50.

[18] Collins S. L. Interaction of Disturbances in Tallgrass Prairie: A Field Experiment [J]. Ecology, 1987, 68 (5): 1243-1250.

[19] Collins S. L., J. A. Bradford, P. L. Sims. Succession and Fluctuation in Artemisiadominated Grassland [J]. Vegetatio, 1988, 73 (2): 89-99.

[20] Connell J. H. Diversity in Tropical Rain Forests and Coral Reefs [J]. Science, 1978, 199 (4335): 1302-1310.

[21] Conte T. J., B. Tilt. The Effects of China's Grassland Contract Policy on Pastoralists' Attitudes Towards Cooperation in an Inner Mongolian Banner [J]. Human Ecology, 2014, 42 (6): 837-846.

[22] Coppock D. L., J. E. Ellis, D. M. Swift. Seasonal Patterns of Activity, Travel and Water Intake of Livestock in South Turkana, Kenya [J]. Journal of Arid Environments, 1988, 14 (3): 319-331.

[23] Cornelissen J. H. C., S. Lavorel., E. Garnier. A Handbook of Protocols for Standardised and Easy Measurement of Plant Functional Traits Worldwide [J]. Australian Journal of Botany, 2003, 51 (4): 335-380.

[24] Danell, Lundberg, Hjalten. Herbivore Avoidance by Association: Vole and Hare Utilization of Woody Plants [J]. Oikos, 1993, 68 (1): 125-131.

[25] David, Sneath. Social Relations, Networks and Social Organisation in Post – Socialist Rural Mongolia [M]. Mongolila: Nomadic Peoples, 1993.

[26] Deák B., B. Tóthmérész, O. Valkó. Cultural Monuments and Nature Conservation: A Review of the Role of Kurgans in the Conservation and Restoration of Steppe Vegetation [J]. Biodiversity and Conservation, 2016, 25 (12): 2473 – 2490.

[27] Derner J. D., R. L. Gillen, F. T. McCollum, et al. Little Bluestem Tiller Defoliation Patterns under Continuous and Rotational Grazing [J]. Journal of Range Management, 1994, 47 (3): 220 – 225.

[28] Dickman M. The Effect of Grazing by Tadpoles on the Structure of a Periphyton Community [J]. Ecology, 1968, 49 (6): 1188 – 1190.

[29] Dyer A. R. Burning and Grazing Management in a California Grassland: Effect on Bunchgrass Seed Viability [J]. Restoration Ecology, 2002, 10 (1): 107 – 111.

[30] Dyksterhuis E. J. Condition and Management of Range Land Based on Quantitative Ecology [J]. Journal of Range Management, 1949, 2 (3): 104 – 115.

[31] Du C., M. Shinoda, K. Tachiiri, et al. Mongolian Nerders' Vulnerability to Dzud: A Study of Record Livestock Mortality Leveis during the Severe 2009/2010 Winter [J]. Natural Hazards, 2017 (1): 1 – 15.

[32] Endicott E. Beyond Great Walls: Environment, Identity, and Development on the Chinese Grasslands of Inner Mongolia (Review) [J]. China Review International, 2003, 10 (1): 272 – 277.

[33] Fan L., Y. Gao, H. Brück, et al. Investigating the Relationship Between NDVI and LAI in Semi – Arid Grassland in Inner Mongolia Usingin – Situmeasurements [J]. Theoretical and Applied Climatology, 2009, 95 (1 – 2): 151 – 156.

[34] Fartmann T., B. Kraemer, F. Stelzner. Orthoptera as Ecological Indicators for Succession in Steppe Grassland [J]. Ecological Indicators, 2012 (20): 337 – 344.

[35] Firincioglu H. K. , S. S. Seefeldt, B. Sahin. The Effects of Long – Term Grazing Exclosures on Range Plants in the Central Anatolian Region of Turkey [J]. Environmental Management, 2007, 39 (3): 326 – 337.

[36] Gan L. , X. Peng, S. Peth, et al. Effects of Grazing Intensity on Soil Water Regime and Flux in Inner Mongolia Grassland, China [J]. Pedosphere, 2012, 22 (2): 165 – 177.

[37] Glenn – Lewin D. C. , R. K. Peet, T. T. Veblen. Plant Succession: Theory and Prediction [J]. Journal of Ecology, 1993, 81 (4): 830 – 831.

[38] Godron M. , P. Daget, J. Poissonet. Some Aspects of Heterogeneity in Grasslands of Cantal (France) [C]. International Symposium on Stat Ecology New Haven, 1971.

[39] Goodman, Daniel. The Theory of Diversity – Stability Relationships in Ecology [J]. Quarterly Review of Biology, 1975, 50 (3): 237 – 266.

[40] Griffin G. D. , M. G. Nolan, C. E. Easterly. Concerning Biological Effects of Spark – Decomposed SF/sub 6/ [J]. IEE Proceedings A – Physical Science, Measurement and Instrumentation, Management and Education, 2006, 137 (4): 221 – 227.

[41] Grime J. P. Competitive Exclusion in Herbaceous Vegetation [J]. Nature, 1973, 242 (5396): 344 – 347.

[42] Grime J. P. Benefits of Plant Diversity to Ecosystems: Immediate, Filter and Founder Effects [J]. Journal of Ecology, 1998, 86 (6): 902 – 910.

[43] Han O. , M. E. Ritchie. Effects of Herbivores on Grassland Diversity [J]. Trends in Ecology & Evolution, 1998, 13 (7): 261 – 265.

[44] Hao J. , U. Dickhoefer, L. Lin. Effects of Rotational and Continuous Grazing on Herbage Quality, Feed Intake and Performance of Sheep on a Semi – Arid Arass-

land Steppe [J]. Archives of Animal Nutrition, 2013, 67 (1): 62 – 76.

[45] Hart R. H. Smart: A Simple Model to Assess Range Technology [J]. Journal of Range Management, 1989, 42 (5): 421 – 424.

[46] Heitschmidt R. K., S. L. Dowhower, J. W. Walker. Some Effects of a Rotational Grazing Treatment on Quantity and Quality of Available Forage and Amount of Ground Litter [J]. Journal of Range Management, 1987, 40 (4): 318 – 321.

[47] Heitschmidt R. K., R. A. Gordon, J. S. Bluntzer. Short Duration Grazing at the Texas Experimental Range: Effects on Forage Quality [J]. Journal of Range Management, 1982, 35 (3): 372 – 374.

[48] Huete A. R. A Soil – Adjusted Vegetation Index (SAVI) [J]. Remote Sensing of Environment, 1988, 25 (3): 295 – 309.

[49] Humphrey R. R. Fire in the Deserts and Desert Grassland of North America [J]. Fire and Ecosystems, 1974: 365 – 400.

[50] Hurlbert S. H. The Nonconcept of Species Diversity: A Critique and Alternative Parameters [J]. Ecology, 1971, 52 (4): 577 – 586.

[51] Jacobo E. J., A. M. Rodríguez, N. Bartoloni, et al. Rotational Grazing Effects on Rangeland Vegetation at a Farm Scale [J]. Rangeland Ecology & Management, 2006, 59 (3): 249 – 257.

[52] Kahmen S., P. Poschlod. Plant Functional Trait Responses to Grassland Succession over 25 Years [J]. Journal of Vegetation Science, 2004, 15 (1): 21 – 32.

[53] Kawamura K., T. Akiyama, H. O. Yokota, et al. Quantifying Grazing Intensities Using Geographic Information Systems and Satellite Remote Sensing in the Xilingol Steppe Region, Inner Mongolia, China [J]. Agriculture Ecosystems & Environment, 2005, 107 (1): 83 – 93.

[54] Kleyer M. Validation of Plant Functional Types Across Two Contrasting

Landscapes [J]. Journal of Vegetation Science, 2002, 13 (2): 167 – 178.

[55] Koch M., B. Schroder, A. Gunther, et al. Taxonomic and Functional Vegetation Changes after Shifting Management from Traditional Herding to Fenced Grazing in Temperate Grassland Communities [J]. Applied Vegetation Science, 2017, 20 (2): 259 – 270.

[56] Kraft N. J. B., L. S. Comita, J. M. Chase. Disentangling the Drivers of? Diversity Along Latitudinal and Elevational Gradients [J]. Science, 2011, 333 (6050): 1755 – 1758.

[57] Lavorel S., E. Garnier. Predicting Changes in Community Composition and Ecosystem Functioning from Plant Praits: Revisiting the Holy Grail [J]. Functional Ecology, 2002, 16 (5): 545 – 556.

[58] Lavorel S., S. Mcintyre, J. J. Landsberg. Plant Functional Classifications: From General Groups to Specific Groups Based on Response to Disturbance [J]. Trends in Ecology & Evolution, 1997, 12 (12): 474 – 478.

[59] Li J., G. S. Okin, L. Alvarez, et al. Quantitative Effects of Vegetation Cover on Wind Erosion and Soil Nutrient Loss in a Desert Grassland of Southern New Mexico, Usa [J]. Biogeochemistry, 2007, 85 (3): 317 – 332.

[60] Li S., P. H. Verburg, S. Lv, et al. Spatial Analysis of the Driving Factors of Grassland Degradation under Conditions of Climate Change and Intensive Use in Inner Mongolia, China [J]. Regional Environmental Change, 2012, 12 (3): 461 – 474.

[61] Loeser M. R. R., T. D. Sisk, T. E. Crews Impact of Grazing Intensity during Drought in an Arizona Grassland [J]. Conservation Biology, 2007, 21 (1): 87 – 97.

[62] Lorenzo P., E. Pazos – Malvido, M. Rubido – Bará, et al. Invasion by the Leguminous Tree Acacia Dealbata (Mimosaceae) Reduces the Native Understorey Plant

Species in Different Communities [J]. Australian Journal of Botany, 2012, 60 (8): 669.

[63] Ma Feng Yun. Research Advances on Ecosystem Stability [J]. Journal of Desert Research, 2002, 22 (4): 401-407.

[64] Marcelo S., M. Gutman, A. Perevolotsky, et al. Vegetation Response to Grazing Management in a Mediterranean Herbaceous Community: A Functional Group Approach [J]. Journal of Applied Ecology, 2000, 37 (2): 224-237.

[65] Matthew C., G. Lemaire, N. R. S. Hamilton, et al. A Modified Self-Thinning Equation to Describe Size/Density Relationships for Defoliated Swards [J]. Annals of Botany, 1995, 76 (6): 0-587.

[66] Mccall D. G., D. A. Clark. Optimized Dairy Grazing Systems in the Northeast United States and New Zealand I. System Analysis [J]. Journal of Dairy Science, 1999, 82 (8): 1808-1816.

[67] Mcintyre S., K. M. Heard, T. G. Martin. The Relative Importance of Cattle Grazing in Subtropical Grasslands: Does It Reduce or Enhance Plant Biodiversity? [J]. Journal of Applied Ecology, 2003, 40 (3): 69-72.

[68] Mcintyre S., S. Lavorel. Livestock Grazing in Subtropical Pastures: Steps in the Analysis of Attribute Response and Plant Functional Types [J]. Journal of Ecology, 2001, 89 (2): 209-226.

[69] Mcnaughton D. A. F. J. Stability Increases with Diversity in Plant Communities: Empirical Evidence from the 1988 Yellowstone Drought [J]. Oikos, 1991, 62 (3): 360-362.

[70] Mcnaughton M., J. Oesterhelds. Effect of Stress and Time for Recovery on the Amount of Compensatory Growth after Grazing [J]. Oecologia, 1990, 85 (3): 305-313.

[71] Metera E., T. Sakowski, K. Słoniewski, et al. Grazing as a Tool to Maintain Biodiversity of Grassland - A Review [J]. Animal Science Papers & Reports, 2010, 28 (4): 315-334.

[72] Milchunas D. G., O. E. Sala, W. K. Lauenroth. A Generalized Model of the Effects of Grazing by Large Herbivores on Grassland Community Structure [J]. American Naturalist, 1988, 132 (1): 87-106.

[73] Milchunas D. G., M. W. Vandever. Grazing Effects on Plant Community Succession of Early - and Mid - Seral Seeded Grassland Compared to Shortgrass Steppe [J]. Journal of Vegetation Science, 2014, 25 (1): 22-35.

[74] Milton S. J. A Conceptual Model of Arid Rangeland Degradation - The Escalating Cost of Declining Productivity [J]. BioScience, 1994, 44 (2): 70-76.

[75] Minami K., J. Goudriaan, E. Lantinga, et al. Significance of Grasslands in Emission and Absorption of Greenhouse Gases [C]. Proc. 17th Int. Grassland Congr., Palmerston North, New Zealand, 1993.

[76] Moreno Garcıa C. A., J. Schellberg, F. Ewert, et al. Response of Community - Aggregated Plant Functional Traits along Grazing Gradients: Insights from African Semi - Arid Grasslands [J]. Applied Vegetation Science, 2014 (17): 470-481.

[77] Mouillot D., D. R. Bellwood, C. Baraloto. Rare Species Support Vulnerable Functions in High - Diversity Ecosystems [J]. PLos Biology, 2013, 11 (5): e1001569.

[78] Na Y., J. Li, B. Hoshino, et al. Effects of Different Grazing Systems on Aboveground Biomass and Plant Species Dominance in Typical Chinese and Mongolian Steppes [J]. Sustainability, 2018, 10 (12): 1-14.

[79] Oba G., O. R. Vetaas, N. C. Stenseth. Relationships Between Biomass

and Plant Species Richness in Arid – Zone Grazing Lands [J]. Journal of Applied Ecology, 2001, 38 (4): 836 – 845.

[80] Olff H., M. E. Ritchie. Effects of Herbivores on Grassland Diversity [J]. Trends in Ecology & Evolution, 1998, 13 (7): 261 – 265.

[81] Orr R. J., J. E. Cook, R. A. Champion, et al. Intake Characteristics of Perennial Ryegrass Varieties When Grazed by Sheep under Continuous Stocking Management [J]. Euphytica, 2003, 134 (3): 247 – 260.

[82] Parsons S. S. D. The Savory Grazing Method [J]. Rangelands, 1980, 2 (6): 234 – 237.

[83] Pérez H. N., S. Díaz, E. Garnier. New Handbook for Standardised Measurement of Plant Functional Traits Worldwide [J]. Australian Journal of Botany, 2013, 61 (3): 167 – 234.

[84] Pfister J. A., G. B. Donart, R. D. Pieper, et al. Cattle Diets under Continuous and Four – Pasture, One – Herd Grazing Systems in Southcentral New Mexico [J]. Journal of Range Management, 1984, 37 (1): 50 – 54.

[85] Philipp H. Wan, M. Gierus, et al. Grassland Responses to Grazing: Effects of Grazing Intensity and Management System in an Inner Mongolian Steppe Ecosystem [J]. Plant & Soil, 2011, 340 (1 – 2): 103 – 115.

[86] Pickett S. T. A., J. Kolasa, et al. Ecological Understanding [M]. New York: Elsevier, 1994.

[87] Pulido R. G., J. D. Leaver. Continuous and Rotational Grazing of Dairy Cows – The Interactions of Grazing System with Level of Milk Yield, Sward Height and Concentrate Level [J]. Grass & Forage Science, 2003, 58 (3): 265 – 275.

[88] Richards, Olsonj. Spatial Arrangement of Tiller Replacement in Agropyron Desertorum Following Grazing [J]. Oecologia, 1988, 76 (1): 7 – 10.

[89] Romme W. H. Fire and Landscape Diversity in Subalpine Forests of Yellowstone National Park [J]. Ecological Monographs, 1982, 52 (2): 199-221.

[90] Rosenzweig M. L. Species Diversity in Space and Time [M]. London: Cambridge University Press, 1995.

[91] Ruifrok J. L., F. Postma, H. Olff. Scale Dependent Effects of Grazing and Topographic Heterogeneity on Plant Species Richness in a Dutch Salt Marsh Ecosystem [J]. Applied Vegetation Science, 2014 (17): 615-624.

[92] Sampson A. W. Plant Succession in Relation to Range Management [D]. Washington D. C: usDA, 1919, 89 (12): 3425-3427.

[93] Semple A. T. Grassland Improvement in Africa [J]. Biological Conservation, 1971, 3 (3): 173-180.

[94] Severson M. K. E. Vegetation Response to the Santa Rita Grazing System [J]. Journal of Range Management, 1988, 41 (4): 291-295.

[95] Shvidenko A., S. Nilsson. A Synthesis of the Impact of Russian Forests on the Global Carbon Budget for 1961-1998 [J]. Tellus, 2010, 55 (2): 25.

[96] Sternberg M., M. Gutman, A. Perevolotsky. Vegetation Response to Grazing Management in a Mediterranean Herbaceous Community: A Functional Group Approach [J]. Journal of Applied Ecology, 2000, 37 (2): 224-237.

[97] Sternberg T., R. Tsolmon, N. Middleton, et al. Tracking Desertification on the Mongolian Steppe through NDVI and Field-Survey Data [J]. International Journal of Digital Earth, 2011, 4 (1): 50-64.

[98] Su Y. Z., Y. L. Li, J. Y. Cui. Influences of Continuous Grazing and Livestock Exclusion on Soil Properties in a Degraded Sandy Grassland, Inner Mongolia, Northern China [J]. Catena, 2005, 59 (3): 267-278.

[99] Suttie J. M., S. G. Reynolds, C. Batello. Grassland of the World [M].

Beijing: China Agriculture Press, 2011.

[100] Tilman D. Diversity and Productivity in a Long – Term Grassland Experiment [J]. Science, 2001, 294 (5543): 843 – 845.

[101] Tilman D., A. Haddi. Drought and Biodiversity in Grasslands [J]. Oecologia, 1992, 89 (2): 257 – 264.

[102] Timofeev I. V., N. E. Kosheleva, N. S. Kasimov, et al. Geochemical Transformation of Soil Cover in Copper – Molybdenum Mining Areas (Erdenet, Mongolia) [J]. Journal of Soils & Sediments, 2016, 16 (4): 1225 – 1237.

[103] Tachiiri K., M. Shinoda, B. Klinkenberg, et al. Assessing Mongolian Snow Disaster Risk Using Livestock and Satellite Date [J]. Journal of Arid Environments, 2008, 72 (12): 2251 – 2263.

[104] Valencia E., F. T. Maestre, B. Y. Le. Functional Diversity Enhances the Resistance of Ecosystem Multifunctionality to Aridity in Mediterranean Drylands [J]. New Phytologist, 2015, 206 (2): 660 – 671.

[105] Waldhardt R., A. Otte. Indicators of Plant Species and Community Diversity in Grasslands [J]. Agriculture Ecosystems & Environment, 2003, 98 (1 – 3): 339 – 351.

[106] Wang J., D. G. Brown, J. Chen. Drivers of the Dynamics in Net Primary Productivity Across Ecological Zones on the Mongolian Plateau [J]. Landscape Ecology, 2013, 28 (4): 725 – 739.

[107] Wang L., D. Wang, Y. Bai, et al. Spatially Complex Neighboring Relationships among Grassland Plant Species as an Effective Mechanism of Defense Against Herbivory [J]. Oecologia, 2010, 164 (1): 193 – 200.

[108] Wario H. T., H. G. Roba, B. Kaufmann. Responding to Mobility Constraints: Recent Shifts in Resource Use Practices and Herding Strategies in the Borana

Pastoral System, Southern Ethiopia [J]. Journal of Arid Environments, 2016 (127): 222 - 234.

[109] Westoby M., B. Walker, I. Noy - Meir. Opportunistic Management for Rangelands Not at Equilibrium [J]. Journal of Range Management, 1989, 42 (4): 266 - 274.

[110] Wyatt - Smith J. The Purpose of Forests: Follies of Development by Jack Westoby [J]. The Commonwealth Forestry Review, 1987, 68 (4): 318 - 319.

[111] Xiaoli B. I., W. Hong, W. U. Chengzhen, et al. Reasearch on the Bio - Diversity and Stability of the Rare Plant Communities [J]. Journal of Fujian College of Forestry, 2003, 23 (4): 301 - 304.

[112] Zhang M. D. A., E. Borjigin, H. Zhang. Mongolian Nomadic Culture and Ecological Culture: On the Ecological Reconstruction in the Agro - Pastoral Mosaic Zone in Northern China [J]. Ecological Economics, 2007, 62 (1): 19 - 26.

[113] Aruna. Investigation on the Use of Mongolian Language in Balinzuoqi [J]. Chinese Mongolian Studies, 2017 (2): 18 - 25. [阿如娜. 巴林左旗蒙古语言使用状况调查 [J]. 中国蒙古学, 2017 (2): 18 - 25.]

[114] An Yuan, Li Bo, Yang Chi, et al. Stipa Grandis Grassland Productivity and Utilization in Inner Mongolia: I Dynamics of Standing Crop of Pastures in Grazing System [J]. Acta Prataculturae Sinica, 2001 (2): 23 - 28. [安渊, 李博, 杨持, 等. 内蒙古大针茅草原草地生产力及其可持续利用研究Ⅰ. 放牧系统植物地上现存量动态研究 [J]. 草业学报, 2001 (2): 23 - 28.]

[115] An Yuan, Li Bo, Yang Chi, et al. Plant Compensatory Growth and Grassland Sustainable Use [J]. Grassland of China, 2001, 23 (6): 1 - 5. [安渊, 李博, 杨持, 等. 植物补偿性生长与草地可持续利用研究 [J]. 中国草地学报, 2001, 23 (6): 1 - 5.]

[116] Ao Renqi. Change and Innovation of Grassland Property Right System [J]. Inner Mongolia Social Sciences (Chinese version), 2003, 24 (4): 116 – 120. [敖仁其. 草原产权制度变迁与创新 [J]. 内蒙古社会科学（汉文版），2003, 24 (4): 116 – 120.]

[117] Ao Renqi, Dalintai. Research on Sustainable Development of Grassland Pastoral Areas [J]. Journal of Inner Mongolia University of Finance and Economics, 2005 (2): 26 – 29. [敖仁其，达林太. 草原牧区可持续发展问题研究 [J]. 内蒙古财经学院学报，2005 (2): 26 – 29.]

[118] Batunacun, Hu Yunfeng, Biligejifu, et al. Spatial Distribution and Variety of Grass Species on the Ulan Bator – Xilinhot Transect of Mongolian Plateau [J]. Journal of Natural Resources, 2015, 30 (1): 24 – 36. [巴图娜存，胡云锋，毕力格吉夫，等. 蒙古高原乌兰巴托—锡林浩特草地样带植物物种的空间分布 [J]. 自然资源学报，2015, 30 (1): 24 – 36.]

[119] Bai Yongfei, Xu Zhixin, Li Dexin. On the Small Scale Spatial Heterogeneity of Soil Moisture, Carbon and Nitrogen in Stipa Communities of the Inner Mongolia Plateau [J]. Act a Ecologica Sinica, 2002, 22 (8): 1215 – 1223. [白永飞，许志信，李德新. 内蒙古高原针茅草原群落土壤水分和碳、氮分布的小尺度空间异质性 [J]. 生态学报，2002, 22 (8): 1215 – 1223.]

[120] Bao Gang, Bao Yulong, Alateng Tuya, et al. Spatio – temporal Dynamics of Vegetatin Phenology in the Mongolian Plateau during 1982 ~ 2011 [J]. Remote Sensing Technology and Application, 2017, 32 (5): 866 – 874. [包刚，包玉龙，阿拉腾图娅，等. 1982~2011年蒙古高原植被物候时空动态变化 [J]. 遥感技术与应用，2017, 32 (5): 866 – 874.]

[121] Bao Xiuxia, Yi Jin, Liu Shurun, et al. Effects of Different Grazing System on Soil Seed Bank in Typical Steppe of Mongolian Plateau [J]. Chinese Journal of

Grassland, 2020 (5): 68 - 74. [包秀霞,易津,刘书润,等. 不同放牧方式对蒙古高原典型草原土壤种子库的影响 [J]. 中国草地学报, 2020 (5): 68 - 74.]

[122] Bao Yuhai. Study on Land Use Change in Inner Mongolia [D]. Beijing: Institute of Remote Sensing Application Technology, Chinese Academy of Sciences, 1999. [包玉海. 内蒙古土地利用变化研究 [D]. 北京:中国科学院遥感应用技术研究所, 1999.]

[123] Bao Yushan. The Degeneration and Desertification of the Inner Mongolia Grasslands Caud by the Rgulatory System and Policy [J]. Journal of Inner Mongolia Normal University (Philosophy & Social Science), 2003, 32 (3): 28 - 32. [包玉山. 内蒙古草原退化沙化的制度原因及对策建议 [J]. 内蒙古师范大学学报(哲社汉文版), 2003, 32 (3): 28 - 32.]

[124] Bao Yin, Wang Zhongwu, Alata. Effect of Grazing on Soil Moisture of Grassland [J]. Modern Animal Husbandry, 2009 (5): 13 - 15. [宝音,王忠武,阿拉塔. 放牧对草地土壤水分的影响 [J]. 当代畜禽养殖业, 2009 (5): 13 - 15.]

[125] Bilige, Du Shufang. Impact of Population Migration on Grassland Animal Husbandry and Ecological Environment Change in Inner Mongolia [J]. Forward Position, 2016 (9): 100 - 104. [毕力格,杜淑芳. 内蒙古人口迁移对草原畜牧业及生态环境变迁的影响研究 [J]. 前沿, 2016 (9): 100 - 104.]

[126] Cao Xin, Gu Zhihui, Chen Jin, et al. Analysisof Human—Induced Steppe Degradation Based on Remote Sensing in Xilin Gole, Inner Mongolia, China [J]. Journal of Plant Ecology (Formerly Acta Phytoecologica Sinica), 2006, 30 (2): 268 - 277. [曹鑫,辜智慧,陈晋,等. 基于遥感的草原退化人为因素影响趋势分析 [J]. 植物生态学报, 2006, 30 (2): 268 - 277.]

[127] Chang Huining, Li Zhengchun, Liu Rengang, et al. Application and De-

velopment of Grazing Ecology in China [J]. Heilongjiang Animal Husbandry and Veterinary, 1998 (3): 40 - 41. [常会宁, 李正春, 刘仁刚, 等. 放牧生态学方法在我国的应用与发展 [J]. 黑龙江畜牧兽医, 1998 (3): 40 - 41.]

[128] Chang Huining, Xia Jingxin, Xu Zhaohua, et al. Grassland Grazing System and Evaluation [J]. Grassland and Lawn, 1994 (12): 40 - 43. [常会宁, 夏景新, 徐照华, 等. 草地放牧制度及评价 [J]. 草原与草坪, 1994 (12): 40 - 43.]

[129] Cheng Ru, Zhao Nan, Zhao Qingshan. Relationship between Grassland Grazing System and Grassland Productivity [J]. Mechanization of Rural Pastoral Areas, 2014 (3): 29 - 30. [成如, 赵楠, 赵青山. 草地放牧利用制度与草地生产力关系 [J]. 农村牧区机械化, 2014 (3): 29 - 30.]

[130] Cong Rihui, Liu Siqi, Zhu Ling, et al. Study on the Forage - livestock Production and Conversion Efficiency under Short - term Grazing on Typical Steppe [J]. Chinese Journal of Grassland, 2017, 39 (6): 47 - 53. [丛日慧, 刘思齐, 朱羚, 等. 短期放牧下典型草原草畜生产和转化效率研究 [J]. 中国草地学报, 2017, 39 (6): 47 - 53.]

[131] Ding Xiaohui, Gong Li, Wang Dongbo, et al. Grazing Effects on Eco - stoichiometry of Plant and Soil in Hulunbeir, Inner Mogolia [J]. Acta Ecologica Sinica, 2012, 32 (15): 4722 - 4730. [丁小慧, 宫立, 王东波, 等. 放牧对呼伦贝尔草地植物和土壤生态化学计量学特征的影响 [J]. 生态学报, 2012, 32 (15): 4722 - 4730.]

[132] Dong Quanmin, Zhao Xinquan, Ma Yushou, et al. Effects of Yak Grazing Intensity on Quantitative Characteristics of Plant Community in a Two - seasonal Rotation Pasture in Kobresia Parva Meadow [J]. Chinese Journal of Ecology, 2011, 30 (10): 2233 - 2239. [董全民, 赵新全, 马玉寿, 等. 牦牛放牧强度对小嵩草草甸

References

两季轮牧草场植物群落数量特征的影响［J］．生态学杂志，2011，30（10）：2233－2239．］

［133］Fang Jingyun. Global Ecology：Climate Change and Ecological Responses［M］．Beijing：Higher Education Press，2000．［方精云．全球生态学：气候变化与生态响应［M］．北京：高等教育出版社，2000．］

［134］Ge Menghe. On Mongolian Grassland Ecological Culture View［J］．Inner Mongolia Social Sciences，1996（3）：41－45．［格孟和．论蒙古族草原生态文化观［J］．内蒙古社会科学，1996（3）：41－45．］

［135］Geng Wencheng，Ma Ning. Study on Feed Intake of Pregnant Ewes in Rotational Grazing of Artificial Grassland［J］．Chinese Herbivore Science，2000（4）：7－8．［耿文诚，马宁．人工草地划区轮牧考尔木怀孕母羊群体采食量研究［J］．中国草食动物科学，2000（4）：7－8．］

［136］Haishan. Study on Sustainable Development of Farming Pastoral Ecotone in Inner Mongolia［J］．Economic Geography，1995（2）：100－103．［海山．内蒙古农牧交错带可持续发展研究［J］．经济地理，1995（2）：100－103．］

［137］Haishan. Study on the Evolution and Regulation of Man Land Relationship in Inner Mongolia Pastoral Area［M］．Hohhot：Inner Mongolia Education Press，2013．［海山．内蒙古牧区人地关系演变及调控问题研究［M］．呼和浩特：内蒙古教育出版社，2013．］

［138］Han Guanghui. Historical Investigation of Land Desertification in China［J］．Land of China，1999（8）：34－36．［韩光辉．我国土地荒漠化的历史考察［J］．中国土地，1999（8）：34－36．］

［139］Han Guodong. A Comparative Study of Grazing Behaviors of Sheeo in Rotational and Seasonal Continuous Grazing System［J］．Grassland of China，1993（2）：1－4．［韩国栋．划区轮牧和季节连续放牧绵羊的牧食行为［J］．中国草地学

报,1993(2):1-4.]

[140] Han Guodong, Jiao Shuying, Biligetu, et al. Effects of Plant Speciesdiversity and Productivity under Different Stocking Rates in the Stipa Breviflora Griseb. Desert Steppe [J]. Acta Ecologica Sinica, 2005, 27(1): 182-188. [韩国栋,焦树英,毕力格图,等. 短花针茅草原不同载畜率对植物多样性和草地生产力的影响[J]. 生态学报,2005,27(1):182-188.]

[141] Han Guodong, Li Qinfen, Wei Zhijun, et al. Response of Intake and Liveweight of Sheep to Grazing Systems on a Family Ranch Scale [J]. Scientia Agricultura Sinica, 2004(5):124-130. [韩国栋,李勤奋,卫智军,等. 家庭牧场尺度上放牧制度对绵羊摄食和体重的影响[J]. 中国农业科学,2004(5):124-130.]

[142] Han Guodong, Xu Zhixin, Zhang Zutong. A Comparision of Rotation and Continuous-Seasonal Grazing Systems [J]. Journal of Arid Land Resources and Environment, 1990(4):87-95. [韩国栋,许志信,章祖同. 划区轮牧和季节连续放牧的比较研究[J]. 干旱区资源与环境,1990(4):87-95.]

[143] Han Huiguang, Sun Donghui, Chen Guoliang, et al. Experimental Report on Rotational Grazing of Beef Cattle [J]. Heilongjiang Animal Husbandry and Veterinary, 1999(4):12-13. [韩慧光,孙东辉,陈国良等. 划区轮牧饲养肉牛实验报告[J]. 黑龙江畜牧兽医,1999(4):12-13.]

[144] Hao Haiguang. Decision Analysis of Returning Farmland to Forest and Grassland in Horqin Sandy Land Based on Land Suitability Evaluation: A Case Study of Horqin Left Wing Houqi [D]. Hohhot: Master's Thesis of Inner Mongolia Normal University, 2008. [郝海广. 基于土地适宜性评价的科尔沁沙地退耕还林还草决策分析——以科尔沁左翼后旗为例[D]. 呼和浩特:内蒙古师范大学硕士学位论文,2008.]

[145] He Gaoji. Zhifeni and the History of World Conquerors [J]. Ethnic Studies, 1981 (6): 39 – 44. [何高济. 志费尼和《世界征服者史》[J]. 民族研究, 1981 (6): 39 – 44.]

[146] Hou Dongmin. A Fundamental Policy of Efficient Reduction of Population Ecologic Pressure Should Be Clearly Implemented in Grasslands Transformation [J]. Science and Technology Review, 2001. [侯东民. 草原治理应明确贯彻"有效卸载人口生态压力"的基本方针 [J]. 科学导报, 2001.]

[147] Hou FuJiang, Yang Zhongyi. Effects of Grazing of Livestock on Grassland [J]. Acta Ecologica Sinica, 2006, 26 (1): 244 – 264. [侯扶江, 杨中艺. 放牧对草地的作用 [J]. 生态学报, 2006, 26 (1): 244 – 264.]

[148] Hou Xiangyang, Xu Haihong. Research on Carbon Balance of Different Grazing Systems in Stipa Breviflora Desert Steppe [J]. Scientia Agricultura Sinica, 2011, 44 (14): 3007 – 3015. [侯向阳, 徐海红. 不同放牧制度下短花针茅荒漠草原碳平衡研究 [J]. 中国农业科学, 2011, 44 (14): 3007 – 3015.]

[149] Hu Huanyong, Zhang Shanyu. Population Geography of China (Volume I) [M]. Shanghai: East China Normal University Press, 1984. [胡焕庸, 张善余. 《中国人口地理》(上册) [M]. 上海: 华东师范大学出版社, 1984.]

[150] Hu Linghao. The Coordinating Population with Environment, Resource and Economy in the Big Development of West [J]. Qinghai Science and Technology, 2000 (4): 6 – 8. [胡令浩. 西部大开发中人口与环境、资源、经济的协调发展 [J]. 青海科技, 2000 (4): 6 – 8.]

[151] Jia Youling. Grassland Degradation Reasons and Establishment of Grassland Protection Long – term Mechanism [J]. Chinese Journal of Grassland, 2011, 33 (2): 1 – 6. [贾幼陵. 草原退化原因分析和草原保护长效机制的建立 [J]. 中国草地学报, 2011, 33 (2): 1 – 6.]

[152] Jin Guili, Zhu Jinzhong, et al. Discussion on Rangeland Degradation [J]. Grassland and Turf, 2007 (5): 79 – 82. [靳瑰丽,朱进忠,等. 论草地退化 [J]. 草原与草坪, 2007 (5): 79 – 82.]

[153] Li Bo. General Ecology [M] Hohhot: Inner Mongolia University Press, 1993. [李博. 普通生态学 [M]. 呼和浩特: 内蒙古大学出版社, 1993.]

[154] Li Bo. The Rangeland Degradation in North China and Its Preventive Strategy [J]. Scientia Agricultura Sinica, 1997, 30 (6): 1 – 10. [李博. 中国北方草地退化及其防治对策 [J]. 中国农业科学, 1997, 30 (6): 1 – 10.]

[155] Li Jianlong, et al. The Comprehensive Effects of Different Rotational Grazing Intersities (DRGI) on the Soil, Grassland and Sheep Production in a Spring – Autumn Pasture of Sagebruch Desert in the Northern Slope of Tianshan Mountain [J]. RACTA Praticulturae Sinica, 2010, 2 (2): 60 – 65. [李建龙等. 不同轮牧强度对天山北坡低山带蒿属荒漠春秋场土草畜影响研究 [J]. 草业学报, 2010, 2 (2): 60 – 15.]

[156] Li Jinhua, Li Zhenqing, Ren Jizhou. The Effects of Grazing on Grassland Plants [J]. Act a Praticulturae Sinica, 2002, 11 (1): 4 – 11. [李金花,李镇清,任继周. 放牧对草原植物的影响 [J]. 草业学报, 2002, 11 (1): 4 – 11.]

[157] Li Jinxia. The Response of Vegetation – soil – soil Fauna to Desertification in the Western of Ordoos Plateau [D]. Changchun: Northeast Normal University, 2011. [李金霞. 鄂尔多斯高原西部植被—土壤—土壤动物对荒漠化的响应 [D]. 长春: 东北师范大学, 2011.]

[158] Li Qinfen, Han Guodong, Ao Tegen, et al. Approch on Restoration Mechanism of Rotational Grazing System on Desert Steppe [J]. Transactions of the CSAE, 2003 (3): 224 – 227. [李勤奋,韩国栋,敖特根,等. 划区轮牧制度在草地资源可持续利用中的作用研究 [J]. 农业工程学报, 2003 (3): 224 – 227.]

[159] Li Qinfen, Han Guodong, Ao Tegen, et al. Effect of Different Grazing Time on Vegetation in Different Paddocks of the Rotational Grazing Pasture [J]. Chinese Journal of Ecology, 2004 (2): 7 – 10. [李勤奋, 韩国栋, 敖特根, 等. 划区轮牧中不同放牧利用时间对草地植被的影响 [J]. 生态学杂志, 2004 (2): 7 – 10.]

[160] Li Su Ying, Ying Ge, Fu Na, et al. Vegetation Indexes Biomass Models for Typical Semi Arid Steppe—A Case Study for Xilinhot in Northern China [J]. Journal of Plant Ecology (Chinese Version), 2007, 31 (1): 23 – 31. [李素英, 莺歌, 符娜, 等. 基于植被指数的典型草原区生物量模型——以内蒙古锡林浩特市为例 [J]. 植物生态学报, 2007, 31 (1): 23 – 31.]

[161] Li Xiangzhen. Effects of Grazing on Characteristics of Carbon, Nitrogen and Phosphorus Pools in Soil Plant System of Typical Steppe [D]. Beijing: Doctoral Dissertation of Chinese Academy of Sciences, 1999. [李香真. 放牧对典型草原土壤—植物系统中碳、氮、磷库特征的影响 [D]. 北京: 中国科学院博士学位论文, 1999.]

[162] Li Yonghong. Grazing Dynamics of the Species Diversity in Aneurplepidium Chinense Steppe and Stipa Grandis Steppe [J]. Acta Botanica Sinica, 1993, 35 (11): 877 – 884. [李永宏. 放牧影响下羊草草原和大针茅草原植物多样性的变化 [J]. 植物生物学杂志, 1993, 35 (11): 877 – 884.]

[163] Li Yonghong. Research on Grazing Degradation Model of the Main Steppe Rangelands in Inner Mongolia and Some Consideration to Establishing a Computerized Rangeland Montioring System [J]. Chines Bulletin of Botany, 1993 (10): 42 – 51. [李永宏. 内蒙古草原草场放牧退化模式研究及退化监测专家系统雏议 [J]. 植物学通报, 1993 (10): 42 – 51.]

[164] Li Yonghong, Alan. Long Term Effects of Different Grazing Systems on Re-

planted Grassland in Southern New Zealand [J]. Grassland and Lawn, 1995 (1): 22-28. [李永宏,阿兰. 不同放牧体制对新西兰南部补播生草丛草地的长期效应 [J]. 草原与草坪, 1995 (1): 22-28.]

[165] Li Yonghong, Wang Shiping. Response of Plant and Plant Community to Different Stocking Rates [J]. Grassland of China, 1999 (3): 11-19. [李永宏,汪诗平. 放牧对草原植物的影响 [J]. 中国草地学报, 1999 (3): 11-19.]

[166] Lin Huilong, Ren Jizhou. Quantitative Studies of the Effects of Trampling on Typical Steppe of Huanxian in Eastern Gansu, China [J]. Acta Agrestia Sinica, 2008, 16 (1): 97-99. [林慧龙,任继周. 环县典型草原放牧家畜践踏的模拟研究 [J]. 草地学报, 2008, 16 (1): 97-99.]

[167] Liu Jiyuan, Qi Yongqing, Shi Huading, et al. Analysis of Soil Wind Erosion Rate with ~ (137) Cstracer in Taliate Xilinguole Transect of Mongolian Plateau [J]. Science Bulletin, 2007, 52 (23): 87-93. [刘纪远,齐永青,师华定等. 蒙古高原塔里亚特—锡林郭勒样带土壤风蚀速率的 ~ (137) Cs 示踪分析 [J]. 科学通报, 2007, 52 (23): 87-93.]

[168] Liu Jingling, Gao Yubao, He Xingdong, et al. Biodiversity and Stability of the Plant Communities in the Middle and Eastern Parts of Inner Mongolia Steppe [J]. Acta Agrestia Sin Ica, 2006, 14 (4): 390-392. [刘景玲,高玉葆,何兴东,等. 内蒙古中东部草原植物群落物种多样性和稳定性分析 (简报) [J]. 草地学报, 2006, 14 (4): 390-392.]

[169] Liu Taiyong. Study on Rotational Grazing of Artificial Grassland [J]. Prataculture and Animal Husbandry, 1993 (1): 48-51. [刘太勇. 人工草地分区轮牧的研究 [J]. 草业与畜牧, 1993 (1): 48-51.]

[170] Liu Zhenguo. Response of Degraded Grassland in Inner Mongolia to Different Types of Disturbance [D]. Beijing: Institute of Botany, Chinese Academy of Sciences,

2006. [刘振国. 内蒙古退化草原对不同类型干扰的响应研究 [D]. 北京: 中国科学院研究生院 (植物研究所), 2006.]

[171] Liu Zhongkuan, Wang Shiping, Chen Zuozhong. Properties of Soil Nutrients and Plant Community after Rest Grazing in Inner Mongolia Steppe, China [J]. Acta Ecologica Sinica, 2006, 26 (6): 2048 – 2056. [刘忠宽, 汪诗平, 陈佐忠. 不同放牧强度草原休牧后土壤养分和植物群落变化特征 [J]. 生态学报, 2006, 26 (6): 2048 – 2056.]

[172] Liu Zhongling, Wang Wei, Hao Dunyuan, et al. Probes on the Degeneration and Recovery Succession Mechanisms of Inner Mongolia Steppe [J]. Journal of Arid Land Resources and Environment, 2002, 16 (1): 84 – 91. [刘钟龄, 王炜, 郝敦元, 等. 内蒙古草原退化与恢复演替机理的探讨 [J]. 干旱区资源与环境, 2002, 16 (1): 84 – 91.]

[173] Liu jianli, Wang Zongli, Li Ping, et al. Quantitative Anylysis of Relationships among Species of Different Communities in Typical Steppe in Xilingol [J]. Chinese Journal of Grassland, 2013 (5): 76 – 81. [柳剑丽, 王宗礼, 李平, 等. 锡林郭勒典型草原不同群落物种间关系的数量分析 [J]. 中国草地学报, 2013 (5): 76 – 81.]

[174] Lu Nan. The Ideas of Modernization of Shin – Jen Wang from the Economic Medicament of Jirem League [J]. Forward Position, 2018 (3): 109 – 112. [鲁楠. 从《哲盟实剂》看王士仁的近代化思想 [J]. 前沿, 2018 (3): 109 – 112.]

[175] Mao Peisheng. Grassland Science (Fourth Edition) [M]. Beijing: China Agriculture Press, 1982. [毛培胜. 草地学 (第四版) [M]. 北京: 中国农业出版社, 1982.]

[176] Na Yintai, Wulan Tuya, Qin Fuying. Dynamic Monitoring of Horqin Sandy Land Desertification Based on 3S Techniques – A Case Study in Horqin Left Wing Ban-

ner [J]. Journal of Arid Land Resources and Environment, 2010, 24 (10): 53 - 57. [那音太, 乌兰图雅, 秦福莹. 基于3S技术的科尔沁沙地土地荒漠化动态监测——以科尔沁左翼后旗为例 [J]. 干旱区资源与环境, 2010, 24 (10): 53 - 57.]

[177] History Compilation Committee of Animal Husbandry Department of Inner Mongolia Autonomous Region. History of Animal Husbandry Development in Inner Mongolia [M]. Hohhot: Inner Mongolia People's Press, 2000. [内蒙古自治区畜牧厅修志编史委员会. 内蒙古畜牧业发展史 [M]. 呼和浩特: 内蒙古人民出版社, 2000.]

[178] Niu Yujie, Yang Siwei, Wang Guizhen, et al. Relationship Between Plant Species, Life Form, Functional Group Diversity, and Biomass under Grazing Disturbance for Four Years on an Alpine Meadow [J]. Acta Ecologica Sinica, 2018, 38 (13): 4733 - 4743. [牛钰杰, 杨思维, 王贵珍, 等. 放牧干扰下高寒草甸物种、生活型和功能群多样性与生物量的关系 [J]. 生态学报, 2018, 38 (13): 4733 - 4743.]

[179] Peng Qi, Wang Ning. Effects of Different Grazing Systems on Grassland Vegetation [J]. Agricultural Science Research, 2005, 26 (1): 27 - 30. [彭祺, 王宁. 不同放牧制度对草地植被的影响 [J]. 农业科学研究, 2005, 26 (1): 27 - 30.]

[180] Qin Fuying. Vegetation Patterns and Dynamics in Response to Climate Change Across the Mongolian Plateau [D]. Hohhot: Inner Mongolia University, 2019 [秦福莹. 蒙古高原植被时空格局对气候变化的响应研究 [D]. 呼和浩特: 内蒙古大学, 2019]

[181] Qing Shana. Study on Grazing System in Mongolian Plateau [D]. Hohhot: Master's Thesis of Inner Mongolia University, 2014. [庆沙那. 蒙古高原放牧制度研

究 [D]. 呼和浩特：内蒙古大学硕士学位论文，2014.]

[182] Ren Jizhou, Hu Zizhi, Mou Xindai, et al. The Comprehensive Sequential Classification of Grassland and Its Significance in Grassland Genesis [J]. Acta Grassland Sinica, 1980 (1): 12 - 24. [任继周，胡自治，牟新待，等. 草原的综合顺序分类法及其草原发生学意义 [J]. 中国草地学报，1980 (1): 12 - 24.]

[183] Ren Jizhou, Mou Xindai. On Rotational Grazing in Different Areas [J]. Chinese Agricultural Sciences, 1964, 5 (1): 21 - 25. [任继周，牟新待. 试论划区轮牧 [J]. 中国农业科学，1964, 5 (1): 21 - 25.]

[184] Ren Jizhou, Nan Zhibiao, Hao Dunyuan. The Three Major Interfaces within Pratacultural System [J]. Acta Praticulturae Sinica, 2000, 9 (1): 1 - 8. [任继周，南志标，郝敦元. 草业系统中的界面论 [J]. 草业学报，2000, 9 (1): 1 - 8.]

[185] Sanasiqin, Yijin, Bao Xiuxia. Effects of Different Grazing Methods on Vegetation and Soil Characteristics of Desert Steppe in China and Mongolia [J]. Chinese Agronomy Bulletin, 2010, 26 (14): 287 - 291. [萨娜斯琴，易津，包秀霞. 不同放牧方式对中蒙荒漠草原植被和土壤特性的影响 [J]. 中国农学通报，2010, 26 (14): 287 - 291.]

[186] Sa Rengaowa, Ao Tegen, Han Guodong, et al. Comparative Study on Quantitative Characteristics of Plant Communities and Livestock Production Performance in Typical Steppe under Different Grazing Intensities [J]. Grassland and Pratculture, 2010, 22 (4): 47 - 50. [萨仁高娃，敖特根，韩国栋，等. 不同放牧强度下典型草原植物群落数量特征和家畜生产性能的比较研究 [J]. 草原与草业，2010, 22 (4): 47 - 50.]

[187] Shi Yuhui, Zhou Hanxin, Ma Yonglin, et al. Experimental Report on Rotational Grazing of 21000 mu (1974 - 1982) [J]. Sichuan Grassland, 1983 (1):

57 - 62. [施玉辉, 周翰信, 马永林, 等. 二万一千亩划区轮牧试验报告 (1974 ~ 1982) [J]. 四川草原, 1983 (1): 57 - 62.]

[188] Song Naigong. Population of China (Inner Mongolia) [M]. Beijing: China Financial and Economic Press, 1987. [宋乃工. 中国人口 (内蒙古分册) [M]. 北京: 中国财政经济出版社, 1987.]

[189] Sun Shixian, Wei Zhijun, Lv Shijie, et al. Characteristics of Plant Community and Its Functional Groups in Desert Grassland under Effects of Seasonal Regulation of Grazing Intensity [J]. Chinese Journal of Ecology, 2013 (10): 181 - 188. [孙世贤, 卫智军, 吕世杰, 等. 放牧强度季节调控下荒漠草原植物群落与功能群特征 [J]. 生态学杂志, 2013 (10): 181 - 188.]

[190] Tang Xiren. Marco Polo and His Travels [M]. Beijing: Commercial Press, 1981. [唐锡仁. 马可波罗和他的游记 [M]. 北京: 商务印书馆, 1981.]

[191] Wang Shiping, Chen Zuozhong. Response of Sheep Production System to Different Grazing Systems [J]. Acta Grassland Sinica, 1999 (3): 42 - 50. [汪诗平, 陈佐忠. 绵羊生产系统对不同放牧制度的响应 [J]. 中国草地学报, 1999 (3): 42 - 50.]

[192] Wang Shiping, Li Yonghong, Wang Yanfen, et al. Influence of Different Stocking Rates on Plant Diversity of Artemisia Frigida Community in Inner Mongolia Steppe [J]. Acta Botanica Sinica, 2001, 43 (1): 92 - 99. [汪诗平, 李永宏, 王艳芬, 等. 不同放牧率对内蒙古冷蒿草原植物多样性的影响 [J]. 植物学报 (英文版), 2001, 43 (1): 92 - 99.]

[193] Wang Diankui, Liu Zhiwu, Wang De, et al. Investigation and Study on Totational Grazing Method of Dairy Cattle (First Report) —Methods and Measures of Rotational Grazing [J]. Chinese Journal of Animal Husbandry, 1966 (2): 7 - 10

[王殿魁,刘志武,王德,等. 奶牛划区轮牧方法的调查研究（第一报）——轮牧方法与措施［J］. 中国畜牧杂志,1966（2）:7-10.］

［194］Wang Guojie. Effects of Grazing on Grassland Community and Soil Seed Bank in Inner Mongolia and Their Relationship［D］. Beijing:Graduate School of Chinese Academy of Sciences,2005.［王国杰. 放牧对内蒙古草原群落和土壤种子库的影响及两者关系的研究［D］. 北京:中国科学院研究生院（植物研究所）,2005.］

［195］Wang Longgeng, Shen Binhua. A Preliminary Study on the Historical Population of Mongolian Nationality (from the Mid 17th Century to the Mid 20th Century) ［J］. Journal of Inner Mongolia University:Humanities and Social Sciences Edition, 1997（2）:30-41.［王龙耿,沈斌华. 蒙古族历史人口初探（17世纪中叶~20世纪中叶）［J］. 内蒙古大学学报（人文社会科学版）,1997（2）:30-41.］

［196］Wang Shuqiang. A Study on Grazing Pattern and Grazing Internsity on Pasture at Hong Chi Ba［J］. Acta Agrestia Sinica,1995（3）:173-180.［王淑强. 红池坝人工草地放牧方式和放牧强度的研究［J］. 草地学报,1995（3）:173-180.］

［197］Wang Wei, Liu Zhongling, Hao Dunyuan, et al. Research on the Restoring Su Cesion of the Degenerated Grassland in Inner Mongolia I. Basic Characteristics and Driving Force for Restoration of the Degenerated Grassland［J］. Acta Phytoecologica Sinica,1996,20（5）:449-459.［王炜,刘钟龄,郝敦元,等. 内蒙古草原退化群落恢复演替的研究I. 退化草原的基本特征与恢复演替动力［J］. 植物生态学报,1996,20（5）:449-459.］

［198］Wang Xinting, Hou Yali, Liu Fang, et al. Point Pattern Analysis of Dominant Populations in a Degraded Community in Leymus Chinensis + Stipa Grandis Steppe in Inner Mongolia, China［J］. Chinese Journal of Plant Ecology,2011（12）:77-85.

［王鑫厅，侯亚丽，刘芳，等．羊草＋大针茅草原退化群落优势种群空间点格局分析［J］．植物生态学报，2011（12）：77－85．］

［199］Wang Yanfen, Wang Shiping. Influence of Different Stocking Rates on Aboveground Present Biomass and Herbage Quality in Inner Mongolia Steppe［J］．Acta Pratacu Lturae Sinica, 1999（1）：16－21．［王艳芬，汪诗平．不同放牧率对内蒙古典型草原牧草地上现存量和净初级生产力及品质的影响［J］．草业学报，1999（1）：16－21．］

［200］Wang Yongjian, Tao Jianping, Peng Yue. Advances in Species Diversity of Terrestrial Plant Communities［J］．Guihaia, 2006（4）：70－75．［王永健，陶建平，彭月．陆地植物群落物种多样性研究进展［J］．广西植物，2006（4）：70－75．］

［201］Wei Zhijun, Han Guodong, Yang Jing, et al. The Response of Stipa Breviflora Community to Stocking Rate［J］．Grassland of China, 2000（6）：1－5．［卫智军，韩国栋，杨静，等．短花针茅荒漠草原植物群落特征对不同载畜率水平的响应［J］．中国草地学报，2000（6）：1－5．］

［202］Wei Zhijun, Yang Jing, Su Ji'an, et al. A Study on the Standing Forages and Nutrient Dynamics of Community on Stipa Breviflora Grassland under Different Systems［J］．Agricultural Research in the Arid Areas, 2003, 21（4）：53－57．［卫智军，杨静，苏吉安，等．荒漠草原不同放牧制度群落现存量与营养物质动态研究［J］．干旱地区农业研究，2003, 21（4）：53－57．］

［203］Wen Zheng. Influence of Mongolian Plateau Uplift on Asian Climate Formation［J］．Science, 2015（5）：58．［闻正．蒙古高原隆升对亚洲气候形成的影响［J］．科学，2015（5）：58．］

［204］Wulan Tuya. The Development of Agriculture and Changes of Land Ues in Horqin in the 20th Century［J］．Journal of Natural Resources, 2002, 17（2）：157－

161.［乌兰图雅.20世纪科尔沁的农业开发与土地利用变化［J］.自然资源学报，2002，17（2）：157－161.］

[205] Wulan Tuya, Zhang Xueqin. Changes of the Cultivation Norhtern Limit for Horqin Region in the Qing Dynasty ［J］. Scientia Geographica Sinica, 2001, 21 (3): 230－235.［乌兰图雅，张雪芹.清代科尔沁农耕北界的变迁［J］.地理科学，2001，21（3）：230－235.］

[206] Wu Lantuya, Bao Gang, Wu Yundeji, et al. Hyper－spectral Remote Sensing Estimates of Aboveground Biomassof Grassland ［J］. Journal of Mongolia Normal Universiyt (Natural Science Edition), 2015, 44 (5): 660－666.［乌兰图雅，包刚，乌云德吉，等.草地生物量高光谱遥感估算研究［J］.内蒙古师范大学学报（自然科学汉文版），2015，44（5）：660－666.］

[207] Wu Lanbater, Liu Shoudong. A Study on Countermeasures Against Main Climate Damages to Animal Husbandry in Inner Mongolia ［J］. Journal of Natural Disasters, 2004, 13 (6): 36－40.［乌兰巴特尔，刘寿东.内蒙古主要畜牧气象灾害减灾对策研究［J］.自然灾害学报，2004，13（6）：36－40.］

[208] Wulan Tuya, Wu Dun, Na Yintai. The Population Change in Horqin in the 20th Century ［J］. Acta Geographica Sinica, 2007 (4): 418－426.［乌兰图雅，乌敦，那音太.20世纪科尔沁的人口变化及其特征分析［J］.地理学报，2007（4）：418－426.］

[209] Wu Enqi. Comparative Study on Grazing Ecology in Typical Steppe Between China and Mongolia of Mongolian Plateau ［D］. Hohhot: Inner Mongolia Agricultural University, 2017.［吴恩岐.蒙古高原中蒙典型草原放牧生态学比较研究［D］.呼和浩特：内蒙古农业大学，2017.］

[210] Wu Xin, Wu Fangmei, Chen Weimin, et al. Effects of Warm Season Rotational Grazing on Production Performance of Miscanthus Elongatus Type Steppe ［J］.

Heilongjiang Animal Husbandry and Veterinary, 2005 (7): 56 - 57. [武新, 武芳梅, 陈卫民, 等. 暖季轮牧对长芒草型干草原生产性能的影响初报 [J]. 黑龙江畜牧兽医, 2005 (7): 56 - 57.]

[211] Xiji Ritana. The Impact on the Vegetation of Stipa Breviflora Desert Steppe under Different Grazing Systems and Rotational Grazing Time [D]. Hohhot: Inner Mongolia Agricultural University, 2013. [希吉日塔娜. 不同放牧制度和轮牧时间对短花针茅荒漠草原植被的影响 [D]. 呼和浩特: 内蒙古农业大学, 2013.]

[212] Xi Lintuya, Xu Zhu, Zheng Yang. Effects of Grazing on Grassland Plant Community [J]. Caoye Yu Xumu, 2008 (10): 1 - 5, 22. [锡林图雅, 徐柱, 郑阳. 放牧对草地植物群落的影响 [J]. 草业与畜牧, 2008 (10): 1 - 5, 22.]

[213] Xiao Xupei, Song Naiping, Wang Xing, et al. Soil and Water Conservation in China [J]. Soil and Water Conservation in China, 2013 (12): 19 - 23. [肖绪培, 宋乃平, 王兴, 等. 放牧干扰对荒漠草原土壤和植被的影响 [J]. 中国水土保持, 2013 (12): 19 - 23.]

[214] Xie Shuying, Peng Fang. Cause Analysis of Grassland Degradation and Long Term Mechanism of Grassland Protection [J]. Agricultural Development and Equipment, 2017 (9): 55. [谢树瑛, 彭芳. 草原退化原因分析和草原保护长效机制 [J]. 农业开发与装备, 2017 (9): 55.]

[215] Xiong Xiaogang, Han Xingguo, Chen Quansheng, et al. Application of the Equilibrium and Non - equilibrium Ecology to the Dynamics of the Steppe Grazing System in Xilin River Basin, Inner Mongolia [J]. Acta Ecologica Sinica, 2004, 24 (10): 77 - 82. [熊小刚, 韩兴国, 陈全胜, 等. 平衡与非平衡生态学在锡林河流域典型草原放牧系统中的应用 [J]. 生态学报, 2004, 24 (10): 77 - 82.]

[216] Xue Lihong, Cao Weixing, Luo Weihong, et al. Relationship Between Spectral Vegetation Indices and Lai in Rice [J]. Acta Phytoecologica Sinica, 2004

(1): 50 – 55. [薛利红, 曹卫星, 罗卫红, 等. 光谱植被指数与水稻叶面积指数相关性的研究 [J]. 植物生态学报, 2004 (1): 50 – 55.]

[217] Wulantuya. Land Reclamation and Land – use Changes during Last 50 Years in Ke'erqin Deserts, Inner Mongolia [J]. Progress in Geography, 2000, 19 (3): 273 – 278. [乌兰图雅. 科尔沁沙地近 50 年的垦殖与土地利用变化 [J]. 地理科学进展, 2000, 19 (3): 273 – 278.]

[218] Yang Chi. Ecology [M]. Beijing: Higher Education Press, 2008. [杨持. 生态学 [M]. 北京: 高等教育出版社, 2008.]

[219] Yang Dianlin, Han Guodong, Hu Yuegao, et al. Effects of Grazing Intensity on Plant Diversity and Aboveground Biomass of Stipa Baicalensis Grassland [J]. Chinese Journal of Ecology, 2006, 25 (12): 1470 – 1475. [杨殿林, 韩国栋, 胡跃高, 等. 放牧对贝加尔针茅草原群落植物多样性和生产力的影响 [J]. 生态学杂志, 2006, 25 (12): 1470 – 1475.]

[220] Yang Hao, Bai Yongfei, Li Yonghong, et al. Response of Plant Species Composition and Community Structure to Long – Term Grazing in Typical Steppe of Inner Mongolia [J]. Chinese Journal of Plant Ecology, 2009 (3): 79 – 87. [杨浩, 白永飞, 李永宏, 等. 内蒙古典型草原物种组成和群落结构对长期放牧的响应 [J]. 植物生态学报, 2009 (3): 79 – 87.]

[221] Yang Jing, Ba Yinbater. The Use Factor of Several Plant Species in Stipa Breviflora Rangeland [J]. Inner Mongolia Prataculture, 2001 (4): 10 – 14. [杨静, 巴音巴特. 短花针茅草地几种牧草利用率的确定 [J]. 草原与草业, 2001 (4): 10 – 14.]

[222] Yang Jing, Zhu Guilin, Gao Guorong, et al. Effects of Grazing Systems on the Reproductive Feature of Key Plant Population in Stipa Breviflora Steppe [J]. Journal of Arid Land Resources and Environment, 2001 (S1): 112 – 116. [杨静, 朱

桂林，高国荣，等. 放牧制度对短花针茅草原主要植物种群繁殖特征的影响 [J]. 干旱区资源与环境，2001（S1）：112 - 116.］

［223］Yang Limin, Han Mei. Plant Diversity Change in Grassland Communities Along a Grazing Disturbance Gradient in the Northeast China Transect [J]. Acta Phytoecologica Sinica, 2001, 25 (1): 110 - 114. ［杨利民，韩梅. 中国东北样带草地群落放牧干扰植物多样的变化 [J]. 植物生态学报，2001，25（1）：110 - 114.］

［224］Yang Xia, Wang Zhen, Yun Xiangjun, et al. Net Primary Production and Forage Quality of Desert Steppe Plant Communities under Different Grazing Systems and Growing Seasons [J]. Acta Prataculturae Sinica, 2015, 24 (11): 1 - 9. ［杨霞，王珍，运向军，等. 不同降雨年份和放牧方式对荒漠草原初级生产力及营养动态的影响 [J]. 草业学报，2015，24（11）：1 - 9.］

［225］Yang Zhiming, Wang Qin, Wang Xiujuan, et al. Effects of the Different Grazing Intensity on the Phenophase, Viability of Plants and Water Content in Soil [J]. Journal of Agricultural Sciences, 2005 (3): 5 - 7 + 17. ［杨智明，王琴，王秀娟，等. 放牧强度对草地牧草物候期生活力和土壤含水量的影响 [J]. 农业科学研究，2005（3）：5 - 7 + 17.］

［226］Yu Xiao. Analysis of Population Migration and Regional Economic Development in Northeast China since the Founding of the People's Republic of China [J]. Journal of Population, 2006 (3): 29 - 34. ［于潇. 建国以来东北地区人口迁移与区域经济发展分析 [J]. 人口学刊，2006（3）：29 - 34.］

［227］John W. Longworth, Greg J. Williamson. Population Constraints on Pastoral Development in China [J]. China Rural Economy, 1994 (8): 55 - 61. ［约翰·W. 朗沃斯，格里格·J. 威廉森. 中国牧区发展的人口制约因素 [J]. 中国农村经济，1994（8）：55 - 61.］

［228］Yun Heyi. The Main Reasons for Large Amount of Land Reclamation in Inner Mongolia since the Qing Dynasty [J]. Journal of Northern Agriculture, 1999 (6): 12-15. [云和义. 清代以来内蒙古大量开垦土地的主要原因 [J]. 北方农业学报, 1999 (6): 12-15.]

[229] Yun Wenli, Wang Yongli, Liang Cunzhu, et al. Relationship Between Ecohydrological Process and Vegetation Degeneration in Typical Steppe [J]. Chinese Journal of Grassland, 2011, 33 (3): 59-65. [云文丽, 王永利, 梁存柱, 等. 典型草原区生态水文过程与植被退化的关系 [J]. 中国草地学报, 2011, 33 (3): 59-65.]

[230] Zhang Jinping, Zhang Jing, Sun Suyan. Application of Grey Correlation Analysis in Oasis Ecosystem Stability Evaluation [J]. Resources Science, 2006, 28 (4): 195-200. [张金萍, 张静, 孙素艳. 灰色关联分析在绿洲生态稳定性评价中的应用 [J]. 资源科学, 2006, 28 (4): 195-200.]

[231] Zhang Jintun. Quantitative Ecology [M]. Beijing: Science Press, 2004. [张金屯. 数量生态学 [M]. 北京: 科学出版社, 2004.]

[232] Zhang Lu. Effect of Land Use Ways on Functional Traits and Functional Diversity of Advantage Species for the Typical Steppe of Inner Mongolia [D]. Hohhot: Inner Mongolia University, 2011. [张璐. 不同利用方式对内蒙古典型草原优势种功能性状及功能多样性的影响 [D]. 呼和浩特: 内蒙古大学, 2011.]

[233] Zhang Shuhai, Cheng Hong, Li Guisheng, et al. Effects of Different Grazing Methods on Grassland Ecology [J]. Ningxia Agricultural and Forestry Science and Technology, 2006 (6): 24-25. [张树海, 成红, 栗贵生, 等. 不同放牧方式对草原生态影响的试验研究 [J]. 宁夏农林科技, 2006 (6): 24-25.]

[234] Zhang Yaoqing. Travel Notes of Ancient and Modern Times [M]. Beijing: China Press, 1936. [张耀卿. 古今游记丛钞 [M]. 北京: 中华书局, 1936.]

[235] Zhang Zihe. The Effect and Causes of Grassland Degeneration [J]. Pratacultural Science, 1995, 12 (6): 1 – 5. [张自和. 草原退化的后果及深层原因探讨 [J]. 草业科学, 1995, 12 (6): 1 – 5.]

[236] Zhao Dengliang, Liu Zhongling, Yang Gui Xia, et al. Grazing Impact on Distribution Pattern of the Plant Communities and Populations in Stipa Krylovii Steppe [J]. Acta Pratacult Urae Sinica, 2010, 19 (3): 6 – 13. [赵登亮, 刘钟龄, 杨桂霞, 等. 放牧对克氏针茅草原植物群落与种群格局的影响 [J]. 草业学报, 2010, 19 (3): 6 – 13.]

[237] Zhao Gang, Cao Zilong, Li Qingfeng. A Freliminary Study of the Effects of Deferred Spring Grazing on the Pasture Vegetation [J]. Acta Agrestia Sinica, 2003, 11 (2): 183 – 188. [赵钢, 曹子龙, 李青丰. 春季禁牧对内蒙古草原植被的影响 [J]. 草地学报, 2003, 11 (2): 183 – 188.]

[238] Zhao Halin, Zhou Ruilian. Species Source Characters in Restoring Process of Korqin Desertification Grassland Vegetation [J]. Grassland of China, 1994 (4): 1 – 8. [赵哈林, 周瑞莲. 科尔沁沙漠化草场植被恢复过程中的种源特性研究 [J]. 中国草地学报, 1994 (4): 1 – 8.]

[239] Zhao Ji. Effect of Stocking Rates on Soil Microbial Number and Biomass in Steppe [J]. Acta Agrestia Sinica, 1999, 7 (3): 223 – 227. [赵吉. 不同放牧率对冷蒿小禾草草原土壤微生物数量和生物量的影响 [J]. 草地学报, 1999, 7 (3): 223 – 227.]

[240] Zhao Xueyong, Zhou Ruilian, Zhang Tonghui, et al. Boundary Line on Agro – Pasture Zigzag Zone in North China and Its Problems on Eco – Environment [J]. Advance in Earth Sciences, 2002 (5): 110 – 118. [赵学勇, 周瑞莲, 张铜会, 等. 北方农牧交错带的地理界定及其生态问题 [J]. 地球科学进展, 2002 (5): 110 – 118.]

[241] Zhou Hong. Ecosystem and Dissipative Structure [J]. Journal of Ecology, 1989 (4): 51 – 54. [周鸿. 生态系统与耗散结构 [J]. 生态学杂志, 1989 (4): 51 – 54.]

[242] Zhou Xinyin. Impact of Land Use Change and Climate Changeon Vegetation in Mongolian Plateau [D]. Beijing: Beijing Forestry University, 2014. [周锡饮. 气候变化和土地利用对蒙古高原植被覆盖影响 [D]. 北京: 北京林业大学, 2014.]

[243] Zhu Guilin. Comparison Study on the Response of Main Plant Population Character to Different Grazing Systems on Stipa Breviflora Desert Steppe [D]. Hohhot: Inner Mongolia Agricultural University, 2001. [朱桂林. 短花针茅草原主要植物种群特征对不同放牧制度响应的比较研究 [D]. 呼和浩特: 内蒙古农业大学, 2001.]

[244] Zhu Jingfang, Xing Bailing, Ju Weimin, et al. Remote – sensing Estimation of Grassland Vegetation Coverage in Inner Mongolia, China [J]. Chinese Journal of Plant Ecology, 2011, 35 (6): 615 – 622. [朱敬芳, 邢白灵, 居为民, 等. 内蒙古草原植被覆盖度遥感估算 [J]. 植物生态学报, 2011, 35 (6): 615 – 622.]

[245] Zhuo Yi. The Ration Remote Sensing Method Study of Desertification of Mongolia Plateau Based on MODIS Data [D]. Hohhot: Inner Mongolia Normal University, 2007. [卓义. 基于 MODIS 数据的蒙古高原荒漠化遥感定量监测方法研究 [D]. 内蒙古师范大学, 2007.]

[246] Zuo Xiaoan, Zhao Halin, Zhao Xueyong, et al. Spatial Heterogeneity of Soil Organic Carbon and Total Nitrogen of Sandy Grassland in the Restoration of Degraded Vegetation in Horqin Sandy Land, Northern China [J]. Environmental Science, 2009, 30 (8): 2387 – 2393. [左小安, 赵哈林, 赵学勇, 等. 科尔沁沙地退化植被恢复过程中土壤有机碳和全氮的空间异质性 [J]. 环境科学, 2009, 30 (8):

2387-2393.]

[247] Zuo Xiaoan, Zhao Halin, Zhao Xueyong, et al. Species Diversity of Degraded Vegetation in Different Age Restorations in Horqin Sandy Land, Northern China [J]. Acta Praticulturae Sinica, 2009, 18 (4): 9-16. [左小安, 赵哈林, 赵学勇, 等. 科尔沁沙地不同恢复年限退化植被的物种多样性 [J]. 草业学报, 2009, 18 (4): 9-16.]

Appendix

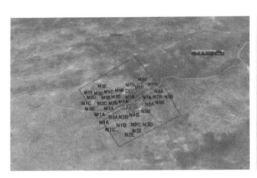

Research area and experimental design

Mongolian grassland

Border grassland

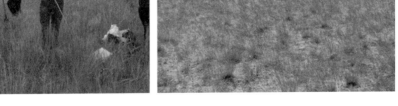

Inner Mongolia grassland

蒙古草原放牧制度效益评价研究
Evaluation Research on the Benefits of Grazing System in the Mongolian Grassland

Dinner on the prairie

Sample survey

Interview with herders

Weigh the vegetation

Horse racing training (Mongolia)

Living conditions in Inner Mongolia

Appendix

Transport of drinking water (Mongolia)

In the herdsmen's house

Grassland without net fence (Mongolia)

Net fence (Inner Mongolia)

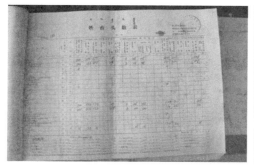

Collect statistics

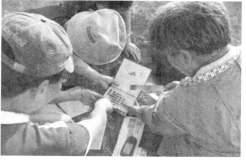

Collect soil data